# Technopreneurship and Sustainability

Technopreneurship is an advanced version of entrepreneurship that incorporates cutting-edge technologies from AI to big data into entrepreneurial activities and leads to innovative solutions to problems that previous solutions have failed to sustain in the long term. This book explains the fundamentals of developing and establishing a small one-person business or a large organization, building technological solutions, and performing effectively and efficiently. It covers all aspects of an entrepreneurial journey, guiding readers from the startup stage, growing a business, sustainability, and clarifies the role of technopreneurship as a tool to enhance sustainability in today's entrepreneurial ventures.

Features

- Explains the theory and the practice of entrepreneurship in today's technological era.
- Presents technological developments, their implementations, and the potential outcomes of technology used in entrepreneurial activities.
- Covers sustainability and the role of technologies to achieve sustainable development goals associated with different ventures.
- Discusses ways to minimize failures of entrepreneurial ventures especially in startups.
- Includes real-life case studies focused on today's problems and solutions.

This book is ideal for undergraduate and graduate students taking courses in Entrepreneurship and Startup as well as business consultants and business owners in any field.

**Higher Education and Sustainability**

J. Paulo Davim, *Professor, Department of Mechanical Engineering, University of Aveiro, Portugal*

This new series fosters information exchange and discussion on higher education for sustainability and related aspects, namely academic staff and student initiatives, campus design for sustainability, curriculum development for sustainability, global green standards: ISO 14000, green computing, green engineering education, index of sustainability, recycling and energy efficiency, strategic sustainable development, sustainability policies, sustainability reports, etc. The series will also provide information on principles, strategies, models, techniques, methodologies, and applications of higher education for sustainability. It aims to communicate the latest developments and thinking as well as the latest research activity relating to higher education, namely engineering education.

*Higher Education and Sustainability: Opportunities and Challenges for Achieving Sustainable Development Goals*
Edited by Ulisses Manuel de Miranda Azeiteiro and J. Paulo Davim

*Designing an Innovative Pedagogy for Sustainable Development in Higher Education*
Edited by Vasiliki Brinia and J. Paulo Davim

*Higher Education: Progress for Management and Engineering*
Edited by Carolina Machado and J. Paulo Davim

*Principles of Entrepreneurship in the Industry 4.0 Era*
Edited by Rajender Kumar, Rahul Sindhwani, Tavishi Tewary, and J. Paulo Davim

*Technopreneurship and Sustainability: Innovation, Challenges, and Opportunities*
Edited by Rajender Kumar, Rahul Sindhwani, Raman Kumar, Punj Lata Singh, and J. Paulo Davim

For more information about this series, please visit: www.crcpress.com/Higher-Education-and-Sustainability/book-series/CRCHIGEDUSUS

# Technopreneurship and Sustainability

## Innovation, Challenges, and Opportunities

Edited by
Rajender Kumar, Rahul Sindhwani, Raman Kumar,
Punj Lata Singh, and J. Paulo Davim

CRC Press
Taylor & Francis Group
Boca Raton London New York

CRC Press is an imprint of the
Taylor & Francis Group, an **informa** business

Designed cover image: Shutterstock

First edition published 2025
by CRC Press
2385 NW Executive Center Drive, Suite 320, Boca Raton FL 33431

and by CRC Press
4 Park Square, Milton Park, Abingdon, Oxon, OX14 4RN

*CRC Press is an imprint of Taylor & Francis Group, LLC*



ISBN: 978-1-032-88350-2 (hbk)
ISBN: 978-1-032-88353-3 (pbk)
ISBN: 978-1-003-53733-5 (ebk)

DOI: 10.1201/9781003537335

Typeset in Times
by Newgen Publishing UK

# Contents

# Preface

New technologies and an increased interest in entrepreneurship are bringing fast-paced change, and with that, promoting the economy as new ideas are transformed into reality. Where these changes meet lies the dynamic field of technopreneurship: technology and entrepreneurship combining together to create new solutions that solve global problems and industries. *Technopreneurship for Ethical and Sustainable Economic Growth: Concepts and Applications* is a comprehensive book that explains and unpacks the term technopreneurship. It shows how one can use its ideas for ethical practices and sustainable development.

In a word, this book was born from the challenge to answer a critical question: How can technology-based entrepreneurship create economic value alongside creating a positive impact for society and the environment? As we move further into the digital age, the importance of integrating ethics, sustainability, and social responsibility into technopreneurial ventures cannot be overstated. This book hopes to guide readers through realizing how technopreneurship can lead to sustainable economic growth despite the challenges of resource deficiencies, environmental damage, and social and economic inequalities.

The chapters in this book are structured to cover all the different aspects of the subject. The first chapter introduces the concept of technopreneurship, including its history, key elements, and subsequent developments. Subsequently, the attention turns to more specific applications of the discipline, as well as the problems and opportunities it poses. Each chapter is expertly written by a contributor who has significant knowledge and practical experience that is relevant to the chosen topic, ensuring that all readers can understand the content.

As Chapter 1 explains the concept of technopreneurship, it clearly states differences between it and regular entrepreneurship. This builds a foundation to help understand how developing technologies act as enablers of future economic growth for enterprises towards solving current world-wide problems by proposing innovative ideas. Chapter 2 revolves around cases of successful companies in technology fields, the two crucial drivers that redefine industries into becoming sustainable.

Chapter 3 shifts focus to leadership strategies and shifts the focus towards female technopreneurs. Transformational and inclusive approaches in leadership are discussed, and the importance of diversity has been emphasized to innovate and make a collaborative culture in the organizations of technology. Chapter 4 discusses the effects that the digital age has had on technopreneurship. Convergence of different technologies has lead towards the creation of new chances for innovation in several other fields.

Chapter 5 introduces the role of digitalization in business growth and development. It provides the reader with insight into how new technologies may impact the business, with a focus on artificial intelligence, big data, and cloud computing on different lines of business. Chapter 6 presents a decision model that helps in solving challenges that technology-driven businesses might face. The chapter guides entrepreneurs on how to manage and survive in this competitive and demanding marketplace.

Chapter 7 discusses the ethical and sustainability problems that come with technopreneurship. It shows ways to include ESG criteria in business models. Chapter 8 highlights the important role of government support and policies in the growth of technopreneurship. It explains how targeted financial incentives, rules, and education are needed for the ecosystem to grow properly.

Finally, Chapter 9 provides a case study analysis of India as a hub of innovation for technopreneurship, particularly focusing on its efforts in Bangalore. It provides detail factors of successes and challenges technopreneurs face and offers insightful lessons for aspiring entrepreneurs as well as policymakers.

This book is for a variety of readers including students, researchers, workers, and policymakers who want to understand and use the power of technopreneurship. Combining ideas from theory with real-world uses, it encourages readers to think carefully about how technology can help create a fair and sustainable future.

I thank the contributors for their skills and hard work that made this book possible. Their efforts show a common goal to improve knowledge and encourage new ideas that help society and the environment. I also thank the readers, whose interest and questions help ideas to grow in this constantly changing field.

As you start this journey into technopreneurship, I hope that this book motivates you to find new opportunities, embrace innovation, and help build a future where technology and business are working together for responsible and long-term economic growth.

# About the Editors

**Rajender Kumar** is presently working as Freelance Auditor and Assessor empaneled with various Certification Bodies to audit the ISO standards implementation. He has auditing experience of more than 200 Man-Days. Previously, he served as an Associate Professor at Manav Rachna International Institute of Research and Studies, Faridabad, India. He is actively involved and associated with various national level societies promoting the Engineering/Technical/Quality concepts in India like IEI (Member Local Chapter Committee), IIIE (Senior Member), QCFI (Governing Council Member), QCI (Registered Lead-Auditor), IIF (Member and Former Honorary Secretary, Faridabad Chapter), ISHRAE etc. He has received the grant from Government of India to conduct activities related to Entrepreneurship Development. He is credited with 60 research papers, one book, and six book chapters. His areas of expertise are Industrial Engineering and Management, Entrepreneurship and Business Incubation, SCM, Lean, Green and Agile Manufacturing Systems.

**Rahul Sindhwani** is a qualified academician with an impressive track record in the field of Operations Management. He is currently serving as an Assistant Professor at the Indian Institute of Management (Sambalpur- Odisha). With over fifteen years of experience in teaching and research, he has established himself as a well-respected authority in his field. Dr. Sindhwani has made significant contributions to the academic community through his research work, which has been published in highly regarded journals. He is also a guest editor of a reputable journal; a testament to his reputation as an accomplished academic. Dr. Sindhwani's expertise lies in the areas of Decision Science, Quantitative Techniques, Sustainability, Operations Management, Supply Chain Management, and Operations Research. He possesses an in-depth understanding of these fields and has contributed to the development of new frameworks and concepts that have helped organizations to improve their operational efficiencies and competitiveness.

**Raman Kumar** is an Assistant Professor in the Department of Mechanical and Production Engineering at Guru Nanak Dev Engineering College, Ludhiana, Punjab, India. He has completed his B.Tech. (Mechanical Engineering), M. Tech. (Production Engineering) and Ph.D. in Mechanical Engineering. He has five years of industry experience, and 20 years of teaching and research experience. His research interest and expertise focus on advanced manufacturing technologies, sustainable machining, optimization techniques, and hybrid renewable energy systems across various materials and processes. He emphasizes integrating machine learning (ML) and artificial intelligence (AI) to optimize machining parameters, enhance material properties, and improve efficiency in industrial and biomedical applications. His expertise extends into renewable energy systems and Industry 4.0., with particular interest in improving energy efficiency, sustainability, and automation within smart manufacturing and industrial contexts. He has featured in more than 150 research publications in national and international conferences, and journals of repute. In addition, he is editor and reviewer of prestigious academic journals. He is a life member of ISTE and SAE India and has received recognition as an ISTE Best Teacher.

**Punj Lata Singh** is a Postdoctoral Fellow in Operations Management at the Indian Institute of Management (IIM), Sambalpur, Odisha. Previously, she served as an Assistant Professor at Amity University, Noida, India. She holds a Ph.D. from Amity University, M.Tech. from MNNIT, Allahabad, and a B.Tech. in Mechanical Engineering from UPTU, Lucknow. With over 13 years of experience in teaching, research, and administration, Dr. Singh is a dedicated lifetime member of the Indian Society of Technical Education. Throughout her academic career, she has advanced both as an educator and researcher in engineering and management. Her areas of interest are operations management, supply chain digitalization, and agile project management. Dr. Singh has actively participated in organizing and attending numerous national and international conferences and has published over 40 research papers in reputable international journals in the areas of Industrial Engineering and Management. She has also contributed as an editor to the book *Multi-Criteria Decision Modelling: Applicational Techniques and Case Studies*, published by CRC Press.

**J. Paulo Davim** received his Ph.D. degree in Mechanical Engineering in 1997, M.Sc. degree in Mechanical Engineering (materials and manufacturing processes) in 1991, Mechanical Engineering degree (5 years) in 1986, from the University of Porto (FEUP), the Aggregate title (Full Habilitation) from the University of Coimbra in 2005, and the D.Sc. from London Metropolitan University in 2013. He is a Senior Chartered Engineer by the Portuguese Institution of Engineers with an MBA and Specialist title in Engineering and Industrial Management. He is also Eur Ing by FEANI-Brussels and Fellow (FIET) by IET-London. Currently, he is a Professor at the Department of Mechanical Engineering of the University of Aveiro, Portugal. He has more than 30 years of teaching and research experience in Manufacturing, Materials, Mechanical, and Industrial Engineering, with special emphasis in Machining and Tribology. He is also interested in Management, Engineering Education, and Higher Education for Sustainability. He has guided large numbers of postdoc, Ph.D. and master's students and has coordinated and participated in several financed research projects and received several scientific awards. He has worked as a project evaluator for ERC-European Research Council and other international research agencies as well as an examiner of Ph.D. theses for many universities in different countries. He is the Editor in Chief of several international journals, Guest Editor of journals, books Editor, book Series Editor and Scientific Advisory for many international journals and conferences. Presently, he is an Editorial Board member of 30 international journals and acts as reviewer for more than 100 prestigious *Web of Science* journals. In addition, he has also published as editor (and co-editor) more than 125 books and as author (and co-author) more than 10 books, 80 book chapters, and 400 articles in journals and conferences.

# Contributors

**Akshay Kumar Annam**
Assistant Professor
SSMRV College
Bangalore

**Ajay Kumar Badhan**
Assistant Professor
Department of Computer Science Engineering
Lovely Professional University | LPU
Punjab, India

**Harish Kumar Banga**
Fashion and Lifestyle Accessory Design Department
NIFT
Kangra, India

**Abhishek Bhattacherjee**
Associate Professor
Department of Computer Science Engineering
Lovely Professional University |LPU
Punjab, India

**Patkar Kalpesh Dilip**
Assistant Professor
School of Law
CHRIST University
India

**Logasakthi Kandasamy**
Associate Professor
School of Business
RV University
Bengaluru, Karnataka

**Swapandeep Kaur**
Assistant Professor
Department of Electrical Engineering
Guru Nanak Dev Engineering College | GNDEC
Ludhiana, India

**Smita Kavatekar**
Assistant Professor
Department of Commerce, School of Commerce, Finance and Accountancy
Christ University
Bangalore

**Ravishankar Krishnan**
Professor
School of Management
Vel Tech Rangarajan Dr Sagunthala R&D Institute of Science and Technology
Chennai, Tamil Nadu

**Arvind Kumar**
Professor of Practice
Sharda School of Business Studies
Sharda University
UP-India

**Rajender Kumar**
Freelance Assessor and Auditor
Adjunct Professor
Department of Mechanical Engineering,
Graphic Era (Deemed to be University), Clement town, Dehradun-248002, India

**Raman Kumar**
Assistant Professor
Mechanical and Production Engineering
Guru Nanak Dev Engineering College
Ludhiana, Punjab, India
and
Jadara Research Center
Jadara University
Irbid, Jordan

**Kiran Kumar M**
Assistant Professor
Faculty of Management – Finance
CMS Business School JAIN (Deemed-to-be) UNIVERSITY
India

**Deepa Kumari**
Associate Professor
Sharda School of Business Studies
Sharda University
UP-India

**Meghna Luthra**
Assistant Professor
Department of Computer Science and Engineering
SET, MRIIRS
Faridabad, India

**Anand Shankar Raja M**
Research Manager
ISBR Business School
Bangalore

**L. Mynavathi**
Associate Professor
Department of Commerce, School of Commerce, Finance and Accountancy
Christ University
Bangalore

**Santhi Narayanan**
Associate Professor
Sharda School of Business Studies
Sharda University
UP-India

**Elantheraiyan Perumal**
Associate Professor
School of Management
Vel Tech Rangarajan Dr Sagunthala R&D Institute of Science and Technology
Chennai, Tamil Nadu

**Rajeev Kumar Saha**
Associate Professor (ME)
J.C. Bose University of Science and Technology, YMCA
Faridabad, India

**Savita Sindhu**
Associate Professor
Department of Computer Science and Engineering
SET, MRIIRS
Faridabad, India

**Rahul Sindhwani**
Assistant Professor
Department of Operations Management
Indian Institute of Management
Sambalpur, Odisha, India

**Kunwar Partap Singh**
Assistant Professor
Department of Electronics and Communication Engineering
Guru Nanak Dev Engineering College
Ludhiana, India

**Punj Lata Singh**
Post Doctoral Fellow
Department of Operations Management
Indian Institute of Management
Sambalpur, Odisha, India

**Sukhpal Singh**
Assistant Professor
Department of Electrical Engineering
Guru Nanak Dev Engineering College
Ludhiana, India

**Karande Suraj Sudam**
Assistant Professor
School of Law
CHRIST University
India

**Sushant**
Assistant Professor
Department of Mechanical Engineering
SET, MRIIRS
Faridabad, India

**Shruti Vashist**
Dean Academics
Manav Rachna University
Faridabad, India

# 1 Technopreneurship
## *Concept and Beyond*

*Rajeev Kumar Saha*

## 1.1 INTRODUCTION

Globalization has created the world as a new market within reach of technopreneurs of any background. Technology is the buzzword today as it drives innovation in products and processes to reach the target customers. Technopreneurs are those entrepreneurs who have technological innovation at the core of their products and technology-based solutions are created as per the needs of the customers. It was in the late 1980s that the word technopreneur was introduced.

Just as entrepreneurship is a process for entrepreneurs to start and oversee their organization, technopreneurship refers to the process of starting and overseeing an organization with an emphasis on technological offerings. Technopreneurship involves a comprehensive understanding of technological advancements, creativity, calculated risk-taking, and business acumen.

Technopreneurial activities play a crucial role in the economic growth of a country. The advancement in technology has widespread into various sectors like energy, communication, computing, manufacturing, space technology, information technology etc. Digital commerce has also further enhanced the reach of innovative technological products to every corner of the world.

In recent times, the digitization of economies has significantly changed the concept of technopreneurship. While technopreneurship was initially linked to software development and online ventures only (Wang et al., 2019), the scope has now expanded to various sectors including education, banking, health, agriculture, etc. This is due to the rapid advancements in technologies like artificial intelligence, machine learning, Internet of Things (IoT), Cloud Computing, Software as a Service (SaaS) etc.

The integration of innovative technologies greatly enhances the processes of the production systems, thereby increasing overall efficiency. Technopreneurs use innovative technologies to improve upon existing products or processes or create new ones for a better overall experience for the users of the products or processes. These innovative technologies may often become the source of a disruptive technology, challenging, disrupting an established product and therefore the related manufacturer. It can be easily understood by the example of data storage devices used in computer systems wherein disruptive technology transformed it from a floppy disk to compact disk to pen drive to cloud storage.

DOI: 10.1201/9781003537335-1

The technological innovation helps in creating a better product thereby increasing the competitiveness of the organization (Olusegun, Akpoviroro, & Adebowale, 2019).

Technopreneurship has greatly enhanced the effectiveness of addressing global issues. The digital advancements have helped in accessing quality education, healthcare, environment, agriculture and many other information expanding its reach to all stakeholders of the society at a global level. It also helps in understanding each ecosystem on our earth thereby making knowledgeable decisions for sustainable development.

While technopreneurship and traditional entrepreneurship, both involve risk-taking, innovation, and the pursuit of profits, there are distinct differences between them. Traditional entrepreneurship typically focuses on conventional business concepts and frameworks, such as manufacturing, retail, or services that are not heavily reliant on cutting-edge technology. In contrast, technopreneurship is intrinsically linked to the development and commercialization of new technologies. Table 1.1 shows the difference between technopreneurship and traditional entrepreneurship.

**TABLE 1.1**
**Difference between technopreneurship and traditional entrepreneurship**

| Basis | Technopreneurship | Traditional Entrepreneurship |
|---|---|---|
| Primary Focus | Technology is the central element of the business model. | Technology may be used to improve operations but is not the central element of the business model. |
| R&D Intensity | The business often revolves around unique, cutting-edge technological solutions that are not widely available in the market, requiring intensive R&D. | The R&D focus can be on improving existing products, optimizing processes, or finding new market niches rather than developing entirely new technologies. |
| Risk and Uncertainty | Often involves higher levels of risk due to the fast-paced nature of technological advancements and the uncertainty of market acceptance. The technology lifecycle can be short, requiring constant innovation to stay competitive. | The risks are often more predictable and manageable, especially if the business operates in established industries with known market dynamics. |
| Scalability and Market Reach | High scalability potential, as digital products and services can often be replicated and distributed with minimal incremental costs. This can lead to rapid growth and global market reach. | In particular, scaling might be more difficult and resource-intensive for companies that depend on tangible goods or regional services. Significant capital expenditures in personnel, inventory, and infrastructure may be necessary for growth. |

**TABLE 1.1 (Continued)**
**Difference between technopreneurship and traditional entrepreneurship**

| Basis | Technopreneurship | Traditional Entrepreneurship |
|---|---|---|
| Capital and Funding | Brings in a lot of venture capital and angel funding since IT companies have such great development potential. Research and development, product creation, and market expansion frequently require funding rounds. | May rely on more conventional finance sources including bank loans, personal savings, and subsidies for small enterprises. Venture finance is still an option, albeit less common, if the company has a significant innovation component. |
| Market Dynamics and Competition | Entails competing in markets that are extremely dynamic and change quickly. Technopreneurs need to keep ahead of the curve in terms of both technology and market trends. | Although competition might be fierce, it can also be steadier and more predictable. Compared to tech-driven industries, changes in the market are typically less frequent and disruptive. |
| Business Models | Frequently incorporates contemporary business concepts such as freemium, platform-based, subscription-based, and SaaS (software as a service) models. Usually, the purpose of these models is to take advantage of the scalability of digital products. | Encompasses a variety of business methods, including as manufacturing, franchising, brick-and-mortar retail, and others. These methods are frequently more conventional and can call for an inventory or physical presence. |

## 1.2 KEY ELEMENTS OF TECHNOPRENEURSHIP

As already discussed, technopreneurship is a fusion of technology and entrepreneurship. To grasp the peculiar opportunities and difficulties in technopreneurship, it is required to understand the underlying key elements. The key elements of technopreneurship are innovation and creativity, technology, risk and uncertainty, unique business strategies, and a supportive ecosystem.

### 1.2.1 Innovation and Creativity

Technopreneurship is fundamentally centred around creativity and innovation. Technopreneurs are frequently motivated by the urge to use new technical developments to address issues or enhance current solutions. Innovation can come in many different forms, like:

- Product innovation is the process of creating completely new goods that fill gaps in the market or enhance already existing ones

- Process innovation is the application of new processes or the enhancement of current ones to improve productivity or quality
- An organization innovation model is the process of coming up with novel approaches to providing clients with value, frequently via utilizing digital platforms or brand-new, tech-driven company models.

According to Choi, Iftitah and Angela (2024), Innovation plays a pivotal role in bridging the chasm between scholarly and hands-on learning, thereby enhancing individuals' capacity to utilize acquired competencies in authentic technical and entrepreneurial endeavours. Their research offers valuable insights into the noteworthy impact of entrepreneurial education, work experience, and technology readiness on the enhancement of technopreneurial competencies.

Ries (2011) talks about a new approach to creating continuous innovation called the Lean Startup.

The approach to innovation can be sustaining innovation (by offering new products that are variants of their core products) or disruptive innovation (by introducing new products into new markets with new customers).

Creativity is essential in generating ideas and finding novel applications for technology. Technopreneurs often need to think outside the box, identifying unique opportunities and envisioning new possibilities that others may overlook.

### 1.2.2 Technology

Technology is the cornerstone of technopreneurship. It encompasses the tools, systems, and knowledge that enable technopreneurs to create innovative products or services. Various technical aspects may be taught through training to effectively cultivate technopreneurial spirit among future technopreneurs (Awaluddin et al., 2024).

The technology component may consist of:

- Software programs, platforms, and systems that offer fixes for certain issues or expedite procedures
- Hardware refers to actual objects, machinery, or parts that use cutting-edge technical innovations
- Advancements in biological and medicinal domains, including bioinformatics and genetic engineering.

Digital platforms are online networks that make it easier to communicate, transact, or provide services. To comprehend and develop these technologies, technopreneurs need to have technical expertise, or be able to acquire it. To stay inventive and competitive, they also need to keep up with technology trends and improvements.

### 1.2.3 Risk and Uncertainty

Technopreneurship entails handling significant levels of risk and uncertainty. Among the risk factors are:

- Technological Risk: This emanates from the chance that a newly developed technology won't function as planned or that a better substitute will emerge
- Market Risk: Is a risk concerned with the uncertainty over market size, customer acceptance, and the capacity to gain a sizable enough part of the market
- Financial Risk: This arises when there is difficulty in managing expenses and cash flow while still obtaining funds for development and expansion
- Regulatory Risk: This emerges due to non-adherence to legal and regulatory frameworks, which may be intricate and diverse among jurisdictions, particularly concerning domains like data privacy, intellectual property, and healthcare.

Technopreneurs must develop strategies to manage risks, like, thorough market research, agile development methodologies, and prudent financial planning.

The organization would then be able to focus on customer value-contributed activities and operational value-added activities by standardizing deployment settings across products and eliminating non-value-added tasks. By doing this, waste will be eliminated, and operating costs will be greatly reduced. Simultaneously, automation will guarantee rapid deployments, maximizing commissioning times overall.

### 1.2.4 Business Models

Technopreneurial ventures often adopt unique business models tailored to technology-based offerings (Ahamat, 2022). Some common business models followed in technopreneurship include:

- Software as a Service (SaaS): Providing software applications on a subscription basis, often hosted in the cloud (Bhardwaj, 2021)
- Freemium: The practice of providing a basic product or service without charge but charging for upgraded features or greater capabilities
- Platform-Based Models: Creating digital platforms that connect users, such as marketplaces, social networks, or content-sharing sites
- E-commerce: The online sale of products and services with the use of digital technologies for logistics and transaction processing
- Subscription Models: Getting clients to pay a regular amount in exchange for ongoing use of a good or service.

These models often allow for scalability and recurring revenue streams, which are attractive features for investors and technopreneurs alike.

### 1.2.5 The Ecosystem and Networking

Technopreneurs need a supportive ecosystem to succeed. This ecosystem consists of:

- Incubators and accelerators: Establishments that support the expansion of start-ups with finance, resources, and mentorship

- Angel and venture capitalists: Important sources of investment for operations expansion and research and development
- Academic and research Institutions: Partners in creativity and technical advancement
- Government and regulatory bodies: Organizations that issue grants, establish industry guidelines, and implement rules.

Technopreneurs can obtain resources, expertise, and alliances necessary for success by networking inside this ecosystem.

## 1.3 THE TECHNOPRENEURIAL PROCESS

The phases and tasks involved in starting and expanding a technology-based company are all included in the technopreneurial process. This procedure entails opportunity identification, product development, capital acquisition, business launch, and business scaling. A combination of technical know-how, business acumen, creativity, and strategic planning is needed at every level. Figure 1.1 depicts the entire technopreneurial process and is discussed further.

### 1.3.1 Idea Generation and Opportunity Recognition

In the technopreneurial process, coming up with ideas and identifying possibilities is the initial step. Finding a workable business idea based on technological innovation requires this step. The following actions need to be taken:

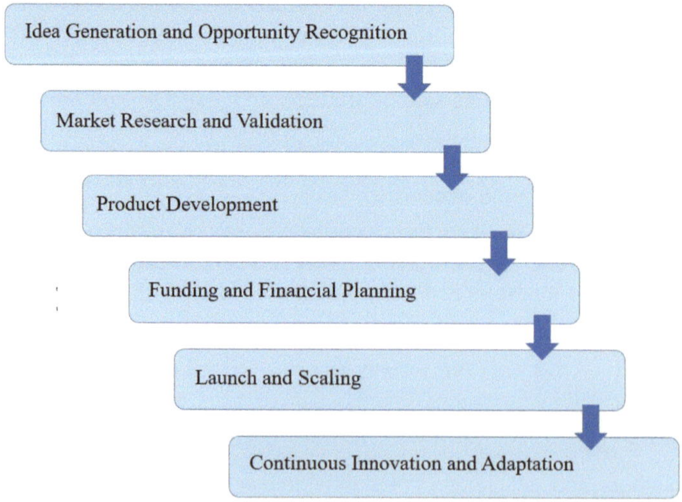


**FIGURE 1.1** Technopreneurial process.

- Technological innovation generally takes place to fulfil the existing needs or Issues in an existing product or process. To find out the needs or issues to be addressed with technological innovation, we have to examine the market or discover any gaps or inefficiencies.
- Once the need or issue to be resolved has been identified, brainstorming is to be done to generate a range of ideas for innovative solutions. However trivial an idea may seem, it should be included for potential solutions. Technical experts should then go through the potential solutions to identify and propose the best feasible ones.
- After identifying the best solutions, their viability should be evaluated based on the technical feasibility and the market potential of the ideas, in terms of the market size, competitors, and current situation.

Technopreneurs who have been successful have often focused on discovering original solutions or some significant advancements to the solutions of the problems (Phuthong, 2023).

### 1.3.2 Market Research and Validation

Assessment of the technological solution coming out of the idea is crucial to validate the market to minimize risks and ensure that the product fulfils an actual market demand. This step further requires following actions.

- Gather data on demand, competitors, target customers, and market trends through methods such as questionnaires, interviews, and market data research. Perform data analysis as part of market research and ensure from the result whether the market is ready to accept the innovative solution.
- The early adopters of an innovative solution are generally sceptical and are more forthcoming with a simplified solution rather than a complex one. The innovative solution should be easy to handle, operate and contain only the features necessary to meet the primary needs of early adopters
- The innovative solution need testing with actual users to identify any missing part or improvement required. This can be done by distributing solution to small group of individuals, collecting their feedback, and continually refining the product. The purpose is to facilitate product enhancement and meet expectations about the needs and preferences of the target market.

### 1.3.3 Product Development

Once the final version of the product is validated by actual users in the marketplace, full-scale product development is done which requires close collaboration between product developers, designers, and engineers with the backing of a strong technical team.

The product development entails the following actions:

- The engineering and development of the product after its detailed designs and specifications are created.

- A prototype of the product to test its functionality and usability and further iterations to improve upon.
- Quality assurance and testing of the product to identify and resolve issues, ensure dependability, and meet quality standards.

### 1.3.4 Funding and Financial Planning

When the product is ready in totality for production, funding is required for procurement of raw material in bulk to make product in larger quantity and sell in the market. Technology-based products usually require significant financial resources to support efforts for product development, marketing, and expansion.

The following actions are required at this stage:

- Examine multiple funding options such as loans, crowdsourcing, venture capital, and angel investments.
- Present a business plan that highlights the company's unique value proposition, market potential, and growth opportunities to persuade possible investors.
- Forecast the financial performance, cash flow management, and budget plan to ensure that the business can stay open and grow.

### 1.3.5 Launch and Scaling

Launching the product into the target market to reach its potential customers requires promotion at various levels. Also, once the demand for the product increases, the supply should be able to match as customers seldom wait. Expanding the company and bringing the product to a larger market requires careful planning so that growth can be managed without compromising client satisfaction or product quality.

The actions required in this phase are:

a) Awareness of product among customers requires a strategic marketing plan to ultimately push the sales of the product. Strategies that can be helpful for the purpose of marketing and sales are partnerships, public relations, and digital marketing.
b) With successful marketing campaigns the demand for the product will grow. The growth will also be seen in the size of the organization. To meet the growing demand for the product and corporate growth, various operations will need to be expanded by hiring more staff, and investing in infrastructure.
c) Customer support systems should be developed to ensure customer care. Customer retention strategies may also be employed such as frequent upgrades and enhancements or loyalty programs and active customer feedback systems.

### 1.3.6 Continuous Innovation and Adaptation

With continuous advancements in technologies, there is a chance of swift shifts in both technology and market conditions. Therefore, innovation and adaptation are necessary all the time to maintain a competitive advantage and ensure long-term prosperity. Technopreneurs are required to constantly innovate and adapt with the help of the following actions:

a) Remaining updated with the latest developments in technology and incorporating new features or improvements to the products.
b) Adjust the strategy by monitoring market shifts and the competitors' activities.
c) Carefully consider user feedback to enhance the product appeal and address any new problems or requests.

## 1.4 CHALLENGES IN TECHNOPRENEURSHIP

Technopreneurship has its own set of challenges that hinders the process of converting an idea into a successful business. The challenges encountered generally stem from the ever-changing nature of technology, unclear markets, regulatory environments, and the need for substantial resources. To succeed in technology driven business environments, challenges must be understood with clarity and overcome with clear strategies.

### 1.4.1 Technological Challenges

- The rapid pace of technological advancements has the potential to develop disruptive technologies which can quickly render the product and services outdated and obsolete. Disruptive technologies being one of the primary challenges, technopreneurs must innovate and adapt to evolving technologies. Staying up to date with new technological developments and trends is essential for technopreneurs who want to remain relevant and competitive.
- To innovate and develop novel technological solutions, specific skills and knowledge is often required. Technopreneurs who want to take on difficult technical problems must possess the necessary technical knowledge or be sure to assemble a capable team who has the knowledge. Many times incorporating technological advancement to innovate the product or service requires highly specialized knowledge, making this stage particularly challenging.
- With access and use of digital technology by more people, it has become more crucial to secure organizations from cyberattacks. Technopreneurs must have robust security procedures and ongoing vigilance to ensure that their products and services are safe from cyberattacks. Financial losses and reputational damage can arise from security vulnerabilities or data breaches.

### 1.4.2 Market Challenges

- Market uncertainty makes it difficult to forecast consumer demand for new technological products. Generally, technopreneurs operate in local regions with limited consumer data regarding preferences and behaviour, predominantly in developing economies. This uncertainty of market demand makes it difficult to plan production, accurately estimate demand, and manage inventories

- At the time of introducing a new technological product or service in the market, it is often necessary to educate potential customers about its features and benefits. The customers are initially reluctant to accept a product or service where the technology is too complex to understand or fundamentally different from existing solutions. Creating awareness about the new technological product or services will help to overcome resistance to change and foster acceptance.
- Numerous firms are vying for market supremacy by evolving new technologies and innovation in the intensely competitive technology sector. Technopreneurs need to differentiate their products from the competitors and build a unique brand in order to stand out. To accomplish this, technopreneurs have to make significant investments in marketing and towards customer acquisition besides focusing on higher-value offerings on their products and services.

### 1.4.3 Regulatory and Legal Challenges

- The legal and regulatory obligations for each of the regions where product and services are to be marketed and sold are generally different and complying with all of them is challenging for the technopreneur. Industry standards, data protection, and intellectual property laws are just a few of the legal obligations that technopreneurs need to follow. Any change in regulations can be costly and time-consuming to comply with, thereby affecting the business operations.
- The innovative technology used by technopreneur to offer a product must be secured from getting copied as it is their most valuable asset. Technopreneur need to protect their intellectual property (IP) as the business is totally dependent on this innovative technology. However, it is very difficult as well as expensive to secure IP (Patents, trademarks, and copyrights etc.). It is equally challenging to prevent infringement of intellectual property rights, particularly in international markets with different legal frameworks.

### 1.4.4 Financial and Resource Challenges

- Funding is required by technopreneurs for research and development, marketing, and further expanding operations on a successful launch and increased demand for the product. It is quite difficult and challenging to get investors especially for young businesses with unproven products or unexplored markets.
- Technopreneurs need to manage finances by controlling the spending, as well as cash flow, and long-term financial planning. Efficient financial management is critical to the survival and growth of technology-related businesses. Through the creation of liquidity issues, financial mismanagement can endanger the venture's success.
- Technology based innovative product can eventually succeed, if a skilled workforce is present to successfully oversee various integrated processes of

the organization. Through hiring and retaining a skilled workforce, specifically in some of the highly competitive fields can be challenging. Technopreneurs should have a compelling vision, competitive compensation packages, and opportunities for career advancement to draw in qualified experts.

### 1.4.5 Strategic and Operational Challenges

- Planning strategically on long-term basis and executing accordingly is essential to succeed in the ever-changing market. Technopreneurs need to adapt to changing circumstances, and strike a balance between immediate operational demands and long-term strategic goals.
- Customer satisfaction is always a top priority which can be achieved by providing a quality product on a continuous basis. To maintain quality, manage increasing demand, and develop infrastructure the organization must overcome operational challenges. Operating inefficiencies can arise from scaling up too rapidly, just as missed opportunities can arise from scaling up too slowly.

## 1.5 SUCCESS FACTORS AND BEST PRACTICES IN TECHNOPRENEURSHIP

Technology is evolving at a rapid pace thereby creating technological innovations and changing the market dynamics. Technopreneur must be able to navigate and understand these changes at the earliest stages to take strategic decisions soonest. Adopting best practices can significantly raise the chances of success even in the face of sometimes challenging circumstances. Few of the best practices worth a mention are strong customer focus, collaborative excellence, networking, ongoing learning and adaptation, outstanding leadership, and sound business strategy. Technopreneurs can more skilfully handle the risks and uncertainties associated with doing business in the technologically advanced world by adhering to these best practices.

### 1.5.1 Strong Leadership and Vision

- A distinct vision and mission statement steers the technopreneurial endeavour, specifically on a long-term basis. A compelling vision statement facilitates a clear path for the company's growth, decision-making support, and group unification. The mission statement describes the company's objectives and the value it hopes to provide to customers and society.
- A strong leader is able to take quick decisions and act strategically to confront adverse situations. As a leader, technopreneurs are required to overcome unforeseen challenges and respond to changing market conditions and technological advancements. Making well-informed decisions is an essential element of proficient leadership, particularly when confronted with ambiguity. Smart leaders cultivate innovative cultures and growth mindsets across the entire organization.

### 1.5.2 Building an Effective Team

- The employees to be selected should have the necessary technical skill sets. They should be go-getters, motivated and informed individuals. Technopreneur should prioritize the selection of individuals who bring diverse perspectives and strengths to the team in the organization. A positive work environment, competitive remuneration, and opportunities for growth and development are all necessary for retaining such talents.
- An atmosphere of transparent communication and cooperation is essential for fostering creativity and resolving issues in a team. Creative thinking and novel solutions can come from interdisciplinary teams of people with different areas of expertise collaborating across functional boundaries. Effective team dynamics also require identifying and resolving disagreements and issues as they arise.

### 1.5.3 Networking and Building Ecosystem Relationships

- After tasting success in initial stages, technopreneurs should actively engage with the wider ecosystem to expand their operations for a broader reach. Engaging a long-lasting relationship with industry experts, incubators, accelerators, and venture capitalists could lead to mentoring, funding, and other advantageous possibilities. Technopreneurs who remain actively involved in the ecosystem can stay current on market trends and opportunities.
- To capture a wider market, technopreneurs should form strategic alliances with other businesses to support and increase operational capabilities. Working in partnership with well-established companies along with research organizations will help access to new markets, clientele, and technology. The alliances will also support co-development, resource sharing, and information exchange further paving the way to reach a wider market with innovative and quality product.

### 1.5.4 Continuous Learning and Adaptation

- Continuous improvement in the organization as a total quality management principle will help in fostering a continuous learning mindset in employees. For this, enhancing knowledge through participation in industry conferences, exhibitions, webinars, publications, and courses is essential. It will also help technopreneurs in knowing the development of new technological innovations and trends and impact on their business.
- Agility to respond quickly to the changes in the market and in customer demands will help in adapting to the changing scenario. Technopreneurs must use agile approaches to make iterative changes in the product to upgrade it as per the requirement of the market. Agility helps to optimize product offerings by matching consumer demands.

### 1.5.5 Customer Focus and Market Understanding

- Understanding target customers' needs to create innovative product requires thorough market research. Technopreneurs should seek feedback from the target customers to continuously improve and enhance the product appeal. Creating products that appeal to the target market can be aided by employing customer-centric strategies like design thinking.
- Customer satisfaction and loyalty is the ultimate goal for a technopreneur to stay in business on a long term. This means creating a brand which is respected and has a brand value associated with the products and services on offer. For creating high brand value, excellent customer offering along with service is necessary to enhance customer satisfaction.

### 1.5.6 Robust Business Strategies

- Customers gets attracted by unique value proposition while investors are attracted to a well-defined business model. A unique value proposition of the product or service on offer emphasizes the unique qualities and differentiates it from other similar products or services. A well-defined business model has clarity on ways to generate revenue and break-even at the earliest.
- Reaching target customers require comprehensive marketing and sales strategies by technopreneurs. The online strategies may include digital marketing, search engine optimization, social media and so on, to reach target customers. While online strategies help in generating substantial leads, it is more important to convert leads into paying customers. Also, data generated through leads can help in campaigning frequently to improve this conversion rate.
- Technopreneurs need to be proficient in financial planning as well as administration for the growth of the organization. Financial planning includes components such as forecasting, budgeting, cash flow management, and cost control. Also, technopreneurs should have different funding sources at their disposal which will be handy when funds may be required to expand the company.

### 1.5.7 Emphasizing Innovation and Differentiation

- Technopreneurs should foster a culture of continuous innovation in order to motivate their team members to experiment with novel ideas and technologies. Regular reviews and enhancements to the company's supply of goods and services are necessary to ensure that it keeps a competitive edge and meets evolving customer demands.
- The product or service which stands out from the rest is crucial for success in a crowded market. Entrepreneurs with innovative business concepts, unique technology, or great user experiences should identify and leverage these strengths. A strong brand and increased market share are facilitated by making a difference.

## 1.6 CASE STUDIES AND EXAMPLES

Successful technopreneurs have overcome challenges with their sheer grit and determination. The challenges faced by technopreneurs and ultimately tasting success by overcoming them can help budding technopreneurs to learn from their case studies and real-life examples. The case studies showcased below will help in understanding the critical decisions, strategies, and adjustments taken by technopreneurs that propel innovative ideas to become lucrative endeavours.

### 1.6.1 Case Study: Apple Inc.

#### 1.6.1.1 Background

Apple Inc., established in 1976 by Ronald Wayne, Steve Jobs, and Steve Wozniak, commonly known for consumer electronics, software, and digital services is one of the best-known examples of technopreneurship. The company has some leading branded products line up specifically iPhone (mobile phone), iPad (tablet), and MacBook (Laptop) by continuously upgrading them.

#### 1.6.1.2 Technopreneurial Journey

Product Development and Innovation: Apple Inc. have differentiated themselves by offering the latest technology with unique designs to attract a niche market. With the release of the iPod, iPhone, and iPad they fundamentally altered their respective markets thanks to their innovative technology, user-friendly interfaces, and exquisite designs.

Business Strategy and Ecosystem: The core of Apple's business strategy is the provision of high-end hardware, complemented by software and services that are developed internally. New revenue streams and a substantial developer community were fostered by the growth of the App Store and iTunes ecosystem.

Difficulties and Adaptation: Apple had to deal with a number of difficulties, such as unsuccessful products (like Newton), shifting leadership, and intense rivalry. The secret to the business's long-term success has been its inventiveness and flexibility in responding to shifting market conditions, especially with the move into mobile and cloud computing.

Key Success Factors: Strong leadership, a focus on innovation and quality, a well-defined ecosystem, and effective branding and marketing strategies.

### 1.6.2 Case Study: Tesla Inc.

#### 1.6.2.1 Background

Tesla Inc. was established in 2003 by Marc Tarpenning and Martin Eberhard. Elon Musk joined in 2004 as its largest shareholder and became CEO of the company. Tesla Inc. is a well-known company in electric vehicles and clean energy. While most of the electric vehicles produced by the company are well received by the customer worldwide, latest model Y became the best-selling electric vehicle in January 2024.

#### 1.6.2.2 Technopreneurial Journey

Innovation and Product Development: Tesla launched its first electric vehicle Roadster sports car in 2008 which was also the first lithium-ion battery-powered electric car on roads. Successive electric vehicle models viz. the Model S, Model 3, and Model X redefined the automobile industry with their technological superiority blended with safety and robust performance.

Business Approach and Ecosystem: Most of the automobile industry in electric vehicle segment face a major problem of providing charging stations to their customers to charge vehicle at different locations. Tesla understood this gap and offered a distinctive value proposition through its massive supercharger network and direct-to-consumer sales approach. On the other hand, the company also branched out into solar products and energy storage systems as part of their strategy towards clean energy.

Difficulties and Adaptation: Tesla had to deal with difficulties such as budgetary limitations, production hold-ups, and problems with quality control. Despite these obstacles, the business has increased output, cut expenses, and turned a profit. Tesla's success is fueled by its capacity to innovate, especially in the areas of autonomous driving and battery technology.

Key Success Factors: Visionary leadership, relentless innovation, vertical integration, and a strong focus on sustainability.

### 1.6.3 Case Study: Zoom Video Communications

#### 1.6.3.1 Background

Eric Yuan founded Zoom Video Communications in 2011, and during the COVID-19 outbreak in particular, the company became well-known. The company's user-friendly platform was quickly embraced by the business, education, and personal communications sectors.

#### 1.6.3.2 Technopreneurial Journey

Innovation and Product Development: Zoom prioritized innovation and product development by ensuring that video conferences were seamless, scalable, and low latency. Features like breakout rooms, virtual backdrops, and strong integration capabilities helped Zoom stand out from the competition.

Business strategy and Scaling: Zoom attracted a healthy user base with its freemium business model and made money from premium subscriptions and enterprise solutions. When demand spiked during the pandemic, the business's architecture enabled quick expansion.

Difficulties and Adaptation: Zoom dealt with "Zoom-bombing", or security and privacy breaches. The company quickly addressed the issue by enhancing user controls and introducing new security features. Taking feedback from the customers on a continuous basis and acting at the earliest wherever necessary with innovative solutions has helped the company to maintain customers' loyalty and trust.

Key Success Factors: A strong business model, scalable infrastructure, user-focused design, and quick problem-solving are essential success factors.

### 1.6.4 Case Study: Spotify

#### 1.6.4.1 Background

Spotify was founded in 2008 by Daniel Ek and Martin Lorentzon to provide access to millions of songs and podcasts through its streaming service. The music business got a makeover with Spotify wherein each individual has access to the music of their choice through their platform.

#### 1.6.4.2 Technopreneurial Journey

Innovation and Product Development: Spotify's platform features personalized playlists, bidirectional streaming between devices, and music recommendation algorithms. Its freemium business model attracted a vast user base from all over the world, offering both a premium subscription and ad-supported free access.

Impact on the Industry and Business concept: Spotify's model completely changed the way music is distributed. It supported musicians by giving them new revenue streams through fan bases. The company has invested in podcasting and original content to broaden its product line and provide content creators with more choices.

Obstacles and Adjustment: Spotify faced stiff challenges for licensing agreements with music copyright owners, competition from major technology companies having similar product line. It also caused a financial burden on getting music licenses as well as the operational expenses of the company. With a focus on original content, constant platform improvements, and expansion into new markets, Spotify has been able to hold the top spot in the music streaming industry.

Key Success Factors: Critical success factors include a creative business plan, solid label relationships, an emphasis on user experience, and deliberate content diversification.

### 1.6.5 Case Study: Airbnb

#### 1.6.5.1 Background

Airbnb was founded in 2008 by Nathan Blecharczyk, Joe Gebbia, and Brian Chesky to let homeowners rent out their houses to travellers through an online platform. It revolutionized the hospitality industry by offering accommodations, a unique alternative to traditional hotels.

#### 1.6.5.2 Technopreneurial Journey

Innovation and Product Development: The online platform created by Airbnb connects hosts and guests with a wide range of lodging options, from private rooms to entire houses. The company has introduced features like verification of hosts as well as guests, host guarantees, and guest reviews to promote safety and confidence within the community.

Business Model and Scaling: The business model of Airbnb includes service charges from both hosts and guests. The business is scalable as it eliminates the requirement for real property ownership. The company has expanded globally by adapting to the local knowledge as well as various regulatory environments.

Difficulties and Adaptation: During the COVID-19 pandemic, there was travel ban almost all over the world, thereby forcing Airbnb to contend with regulations and laws. The company changed by introducing long-term stay options, increasing health and safety measures, and offering hosts flexible cancellation policies.

Key Success Factors: An innovative business strategy, a strong sense of brand identity, an emphasis on trust and community, and flexibility in response to shifting market conditions are critical success factors.

## 1.7 FUTURE TRENDS IN TECHNOPRENEURSHIP

Technopreneurship is bound to see radical transformation due to rapid technological innovation happening in almost every field. It is anticipated that advances in technology development, societal needs, and shifts in market demands in the future will have an impact on technopreneurship trends. Technopreneurs need to be aware of these trends if they hope to stay on top of developments and seize new chances. Future trends that are expected to affect technopreneurship are examined in this section.

### 1.7.1 Artificial Intelligence (AI) and Machine Learning (ML)

- Technopreneurs are utilizing AI as well as ML for better solutions by automating the processes, giving customized and personalized experiences to each of their customers as per needs. AI and ML uses real data with advanced analytics which creates additional avenues for solutions and applications. AI and ML driven innovations are anticipated to impact several industries, like finance, retail, healthcare, and transportation, etc.
- With advancement in AI and ML, various issues related to ethics, transparency, and biasing are bound to crop up. Technopreneurs need to handle ethical concerns and develop responsible AI practices to ensure that their technologies are used securely and fairly.

### 1.7.2 Blockchain and Decentralization

- Blockchain technology being used in cryptocurrencies. However, it is expected to have an impact on various industries such as healthcare, real estate, digital identity, supply chain management, and decentralized finance (DeFi). Technopreneurs are going to explore the potential applications of blockchain technology in various industries to improve productivity, security, and transparency.
- The emergence of semantic web, commonly known as web3, aims to create a more user-centric and decentralized internet. It will drive innovation in several areas, including digital ownership, smart contracts, and decentralized applications. Technopreneurs will get the opportunity to develop and expand decentralized platforms that offer greater privacy and control to the users.

### 1.7.3 Sustainability and Green Technologies

- Global concern for the environment for sustainability of earth will increasingly affect technopreneurship. Eco-friendly products and sustainable practices must be developed by technopreneurs to counteract resource depletion and global warming. Energy efficiency, waste management, and renewable energy innovations are critical to constructing a sustainable future.
- The concept of circular economy is gaining traction which emphasize recycling, reusing, and reducing waste. Technopreneurs must develop cutting-edge business plans with research in futuristic technologies that support the circular economy and lessen environmental impact.

### 1.7.4 Advanced Connectivity and IoT

- With 5G technology already in place and its successor 6G technology being the future standard in wireless communications, resulting in faster and more dependable connectivity, the growth of the internet of things (IoT) and smart devices is imminent. Currently, technopreneurs leverage 5G technology to develop new applications for sectors like manufacturing, healthcare, retail, energy, transport, smart cities etc.
- The usability of IoT is more common now, generating an enormous volume of data. This enormous amount of data needs processing in a planned manner along with the responsibility of its storage and security. Technopreneurs should fully realize the potential of connected technology using innovations in data analytics and IoT integration for smart devices.

### 1.7.5 Personalized and Immersive Experiences

- Augmented Reality (AR) and Virtual reality (VR) are becoming more accessible and immersive with increase in their usability in various sectors including manufacturing, gaming, education, and healthcare. Technopreneurs will create new interactive experiences and applications that use AR and VR to engage users in new ways.
- Advances in artificial intelligence and data analytics will enable highly personalized user experiences. Utilizing data-driven solutions, technopreneurs will develop goods and services that are tailored to the preferences and actions of individual customers, boosting satisfaction and loyalty.

### 1.7.6 Health Tech and Biotech Innovations

- A consequence of the COVID-19 pandemic is the increased demand for telemedicine and remote health solutions. Future trends to watch out for include technological advancements in wearable medical devices, virtual consultations, and remote diagnosis. Tech-savvy businesspeople will focus on leveraging technology to improve the standard and accessibility of healthcare.

- Technological developments in the healthcare domains will drive advances in many ways including medical devices, diagnosis, medicines, biotechnology, gene therapy etc. Technopreneurs will explore innovative uses of healthcare technologies for medical research, treatment, and diagnosis.

### 1.7.7 Workplace Transformation and Remote Work

- Workplaces are transforming from a physical office to a virtual platform where employees join and perform their duties. This kind of platforms and products allow remote collaboration from any part of the world in a timeless manner thereby enhancing productivity as well as employee well-being. Technopreneurs need to focus on innovations in virtual workspaces, team management, and digital communication.
- Digital transformation is the next big thing happening across a range of industries as automation technologies become more extensively employed in business operations. Technopreneurs are creating methods to increase the efficiency of organizations, reduce the need for manual labour, and simplify processes.

### 1.7.8 Regulatory and Policy Developments

- With the advancement in a technology, the concerned regulatory environment also changes. Legal frameworks should adapt according to the changes in technology. In cyber world, the technology standards are rapidly and constantly changing thereby requiring technopreneurs to stay on top of them. It is therefore essential for technopreneurs to stay informed and adhere to evolving regulations.
- Technopreneurs may impact on the concerned technological areas through advocacy and policy influence. Getting in touch with legislators and advocating for favourable policies can help create an environment that is beneficial to technological innovation and entrepreneurial endeavours.

## 1.8 CONCLUSION

A solid grasp of the core concepts, processes, challenges, and adapting best practices of technology entrepreneurship is crucial for individuals aspiring to this career path. Evolving with the times and staying flexible to adjust to shifting market and technological conditions are essential for long-term success. By utilizing innovation and technology, technopreneurs can significantly impact businesses and society by helping to shape their direction.

The major takeaways from the chapter are summarized below:

- Technopreneurship is the process of launching and managing new technology-based businesses by combining innovative technological solutions with strategic business methods to satisfy customer needs.

- While traditional entrepreneurship is the establishment and management of businesses across a range of sectors, technopreneurship involves the use of innovative technology for the development of new goods, services, and business models.
- The ideation, market research, product development, financial planning, market launching, and scaling are the phases of the technopreneurial process, all centred around agility and adaptability to technological advancements.
- Technopreneurs face unique challenges, including complex technological developments, market volatility, regulatory barriers, and scarce funding. In order to overcome these challenges, one needs to be persistent, have a well-thought-out plan, and be able to manoeuvre through a rapidly changing environment.
- The success of a technopreneur requires inclusivity of several best practices, in addition to effective team building, strong leadership, astute networking, and a constant focus on innovation. By using these strategies, technopreneurs can increase their ability to overcome obstacles and seize opportunities.
- A variety of successful paths in technopreneurship are exemplified by actual companies such as Apple, Tesla, Zoom, Spotify, and Airbnb. These case studies highlight the importance of creativity, strategic vision, and adaptability in achieving success and a market impact.
- Technopreneurship will be impacted by future advancements in AI, blockchain technology, sustainability, and remote work, among other trends. Technopreneurs must stay abreast of these developments to modify their strategy, seize new opportunities, and address novel challenges.
- Technopreneurship is defined by a never-ending need for technological innovation and rapid advancements. As technology advances, technopreneurs will become more crucial to the advancement of society, the resolution of international problems, and the direction of numerous industries.
- Technopreneurs must be proactive, adaptable, forward-thinking, and risk taker to explore new markets and technologies. By making use of their resources, connections, and knowledge, they can navigate the complexity of the tech-driven environment and offer significant solutions that promote development and change.

## BIBLIOGRAPHY

Ahamat, A., & Sin, G. K. W. (2022). Developing a business start-up model for technopreneurs. *International Journal of Technoentrepreneurship*, 4(3), 198–218.

Awaluddin, A. I., Komara, E., Rostini, D., & Saifurridjal, A. (2024). Entrepreneurship training model technopreneurship based. *Journal of Research Administration*, 6(1), 7167–7175.

Bhardwaj, B. R. (2021). Adoption, diffusion and consumer behavior in technopreneurship. *International Journal of Emerging Markets*, 16(2), 179–220.

Choi, L. K., Iftitah, N., & Angela, P. (2024). Developing technopreneur skills to face future challenges. *IAIC Transactions on Sustainable Digital Innovation (ITSDI)*, 5(2), 127–135.

Olusegun, O. O., Akpoviroro, K. S., & Adebowale, O. M. (2019). Impact of technopreneurship on business performance. *Journal of Economics and Management Research*, 8, 55–71.

Phuthong, T. (2023). Defining technopreneurs' commercialization research process in the emerging Thai economy: Research for constructing grounded theory. *TEM Journal*, 12(2), 1142–1155.

Reis, E. (2011). *The lean startup*. 1st ed. New York: Crown Business.

Wang, Y. S., Tseng, T. H., Wang, Y. M., & Chu, C. W. (2019). Development and validation of an internet entrepreneurial self-efficacy scale. *Internet Research*, 30(2), 653–675.

# 2 Innovation and Impact
## *Case Studies of Technology-Based Enterprises*

*Sukhpal Singh and Kunwar Partap Singh*

## 2.1 INTRODUCTION

Businesses that use innovative technology to boost operations, disrupt established markets, and provide unique ideas are referred to as technology-based enterprises. These tech-based enterprises integrate progressive technologies like blockchain, internet of things (IoT), and artificial intelligence (AI) into their major processes (Chatterjee et al., 2020). For example, Amazon uses AI and multifaceted algorithms to run its wide-ranging logistical network and provide recommendations for various products to prospective customers (Agrawal et al., 2018). Application-based businesses redefine new limits for the market by innovating new applications tailored for their customer base. For instance, Netflix, through its all-new streaming service, has brought down traditional cable and broadcast television, restructuring the entertainment industry (Lobato, 2019). By creating vital productivity tools and innovative software solutions, businesses like Microsoft and Adobe have revolutionized the software sector (Furman et al., 2019). Moderna and Genentech made groundbreaking engineering advancements in biotechnology and have produced promising treatments for a number of illnesses; for example, Moderna's invention of mRNA technology had a significant role in producing the COVID-19 vaccine (Sahin et al., 2020).

### 2.1.1 Significance of Innovation and Digital Skills

Novelty in digital proficiencies shifts the market scenario, leading to growth in technology-based business. This advancement develops new products and business strategies to satisfy customer needs (Bughin et al., 2017). Moreover, improvement in digitized abilities requires a firm innovative approach. For example, new features and applications in mobile transformed corporate business by opening new business scenarios like in transportation and accommodation. Companies like Uber and Airbnb have uplifted customer satisfaction rates to a great extent. However, mobile applications revolutionize financial handling by providing effortless access to money.

 DOI: 10.1201/9781003537335-2

### 2.1.2 Historical Evolution

Technopreneurship began with some primitive tools and instruments. Personal computer started the beginning of a huge technological revolution in the 1970s. Apple and IBM were leaders in the market at that time which led to redefining personal and corporate development. Further innovation in 1990s and early 2000s led to the advent of internet and digital technology which were critical to the birth of technology leaders of today Amazon and Google Technopreneurship now encompasses a wide range of sectors like block chain, AI, Cloud Computing and advancements in these sectors have opened up new avenues of applications in the sectors of finance, medical, and other industries.

### 2.1.3 Significance of Sustainability in Technopreneurship

Sustainability with reference to business is defined as the ability of the establishment to function in a way so as to guarantee its long term existence and also have a positive effect on society, economy, and mainly the environment. Sustainability in general refers to fulfilling current needs in a way so as to not sacrifice the capacity of future generations and leave ample resources for them to grow further. For instance, the outdoor apparel company Patagonia gives more focus to the use of sustainable materials and production techniques. Ben & Jerry's have certain objectives in their business strategy that are directed towards issues of social justice and climate change. Creating economic value while guaranteeing a just distribution of resources and income is the goal. The resilience and stability of the economy are fostered by a sustainable business model. For example, Unilever's Sustainable Living Plan seeks to increase positive social effect while separating corporate development from environmental impact.

#### 2.1.3.1 Role of Technology in Advancing Sustainable Practices

Technology plays a significant role in emerging sustainable industrial practices with effective use of resources and exhilarating the innovation to provide a sustainable outcome.

How technology promotes sustainability:

- Renewable Energy: Renewable energy sources such as solar and wind power are now much more affordable and efficient. Companies like Tesla and Solar City provide a path with a commitment to deliver solar energy solution by expanding the use of renewable energy and reducing the reliance on fossil fuels
- Energy Efficiency: Smart grids and energy management systems optimize the practice for the usage of energy by lowering the emissions. For example, a big electrical firm like Siemens recuperate the efficiency and reliability of electrical power by incorporating smart grid designs
- Waste Management: Artificial intelligence and IoT tailored the waste outputs from various industrial processes. For example, a waste management system is being employed by a company Rubicon to lower-down the landfill waste by efficient use of resources.

#### 2.1.3.2 Impact on Various Industries

Using technology-driven sustainability methods, operational efficiency has increased, environmental impact has lowered and social responsibility enhanced.

- Agriculture: New age farming technologies like GPS and IoT sensors have significantly increased crop production by helping farmers take into account the use of water, fertilizers and pesticides. Companies such as John Deere have already incorporated these new advancements in their agricultural equipment so as to promote their objective of sustainable agriculture
- Manufacturing: Advancement in the manufacturing methodologies like automation and 3D printing provides a path to uplift the production with optimum use of energy and reduced environmental effects. For instance, General Electric (GE) produces jet engine parts using additive manufacturing, which lowers material waste and increases fuel efficiency
- Transportation: The industry's carbon footprint is being lowered by the introduction of electric vehicles (EVs) and improvements in battery technology. Leading EV manufacturers, such as Tesla and Nissan, provide more environmentally friendly substitutes for conventional gasoline-powered automobiles
- Construction: Energy-efficient materials and smart building systems, among other sustainable building technologies, lessen the environmental effect of construction projects. Leadership in Energy and Environmental Design (LEED) and other green building certifications promote sustainable practices in the construction sector.

### 2.1.4 Objectives of the Chapter

This chapter's main goal is to explore how technology and sustainability interact in the context of tech-based businesses. The objective of this chapter is to illustrate how innovation and digital skills contribute to the development of sustainable business practices by analysing this connection.

- Examine Innovation: This chapter will look at how technical innovation propels tech-based businesses to success. Examining the creation and application of cutting-edge technologies that allow these businesses to disrupt markets and present novel solutions is one aspect of this. For instance, Google's innovations in machine learning and artificial intelligence have transformed a number of industries, including search engines and driverless cars
- Examine Challenges: Financial limitations, regulatory obstacles, and supply chain complexity are just a few of the difficulties that tech-based businesses must overcome. The chapter seeks to identify the resilience and techniques needed to overcome these obstacles by examining them. For example, SpaceX has overcome financial and governmental obstacles to make strides in space exploration
- Identify Opportunities: Technopreneurs may use this chapter to find out how they can use technology to alter society and the environment. This involves investigating the ways in which supporting policy frameworks and stakeholder

participation might promote sustainable innovation. In poor nations, cooperation between digital businesses and non-governmental organisations (NGOs) have the potential to propel sustainable development projects.

This chapter demonstrates the intricate relationship between technology and sustainability in contemporary corporate settings through a number of in-depth case studies. A comprehensive examination of a specific company or project will be provided in each case study, with an emphasis on important components such organisational evolution, market positioning, market dynamics, and technical innovation. The chapter will also examine tech sector failures in order to extract important insights. Knowing the causes of these mistakes might help you identify potential danger spots and tactics that might have produced different results.

This chapter concludes with the intention of providing insightful analysis and motivation for present and future technopreneurs who are dedicated to using technology innovation to have a lasting impact on society.

## 2.2 SELECTION CRITERIA FOR CASE STUDIES

### 2.2.1 Criteria for Choosing Case Studies

Choosing the appropriate case studies is essential to comprehending the complex world of technology-based businesses and how they contribute to sustainability. The case studies included in this chapter were selected based on the following criteria:

- Technological Innovation: Businesses that have developed novel technology or greatly enhanced already-existing ones were chosen. This guarantees that companies at the forefront of technical innovation are highlighted in the case studies. For instance, the automotive and energy storage industries have advanced significantly as a result of Tesla's development of electric automobiles and battery technology
- Market Impact: Companies that have significantly impacted their corresponding markets were selected. This comprises businesses that have completely formed new markets or upended established ones. For example, Beyond Meat's plant-based meat products have completely changed the food sector by encouraging big food chains to adopt more sustainable choices and changing customer behaviour
- Sustainability Efforts: Businesses that actively incorporate sustainability into their business strategies were emphasised. This covers programmes meant to lessen their influence on the environment, encourage social responsibility, and guarantee their economic sustainability. One such example is Fairphone, which is dedicated to producing cellphones that are both ecologically friendly and supplied ethically
- Diverse Sectors: Businesses from a range of industries, such as food, automotive, aerospace, electronics, and biotechnology, were included to give a complete picture. This diversity guarantees that the lessons learned from the case studies may be applied to other sectors

- Success and Failure: A balanced viewpoint is provided by taking into account both successful and unsuccessful businesses. Success stories, like SpaceX's space exploration accomplishments, offer insights on useful tactics and inventions. On the other hand, analysing mistakes—like Theranos' demise—helps pinpoint possible dangers and lessons discovered.

### 2.2.2 Framework for Analysis

A uniform framework is applied to every case study analysis to guarantee a comprehensive and equal analysis of the chosen companies. The following components are part of the framework:

- Market dynamics: An examination of the competitive environment, market size, and client demographics in the context of the business's operations. This aids in comprehending the outside forces affecting the business's performance and strategy
- Market positioning: an analysis of the company's value proposition, competitive advantages, and target market positioning in relation to the market. This draws attention to the strategic choices that influence the success or failure of the business
- Administrative Advancements: Investigating the growth track and prominent variation in the organization structure of business provides a clear insight on timely changes in the business model to meet the demands.
- Technology Advancements for Sustainability: Improvements in the technologies leads to positive impact on the market which in turn improves the performance and sustainability of a business. Moreover assessment of company's financial, social and ecological activities reveals the sustainability of business goals.

This chapter provides a broad insight that how innovative ideas and market analysis creates a pavement for sustainable business.

## 2.3 ANALYTICAL APPROACH

### 2.3.1 Evaluation Methods

The chosen case studies are subjected to a comprehensive analysis using a methodical and multifaceted review process. This method guarantees a thorough comprehension of the tactics, inventions, and sustainability initiatives of any business. Among the techniques of evaluation employed are:

**SWOT Evaluation**

- **Strength:** Analyse various factors like unique and innovative technology, status of company in the market depicts the performance of an establishment. For example, Tesla has potency as its sturdy brand value and exceptional battery performance (Nisar, 2018)

- **Weaknesses:** It means accentuate the inadequacies like volatility in financial matters, inefficient technology and tribulation among employees and management. For illustration, Theranos's employed a mechanism which was not verified and tested trailblazed its collapse (Wilson, 2020)
- Opportunities: Investigating the opportunities for the growth of a company's status due an advancement in the technology and alternations made in the government regulation. For example, plant based diets gained popularity and as compared to based meat based diets
- Threats: Estimating the threats which have negative impact on the accomplishment of goals of the industry or firm, like competition from other major players, recession in the market. In context to this a company Space X faced a strong worldwide competition to achieve its goals.

**PESTLE Evaluation**:

- Political: Investigating the outcomes of political factors with stress on laws, rules, political stability. For instance, Fairphone oversees delicate legal frameworks related to fair trade and E-waste
- Economic: Thorough analysis of the various economic viewpoints which include the various financial circumstances of the current market, different patterns relating to market and economic growth. Teslas' success in the current market can be attributed to the rise of crude prices all over the world and Tesla is successfully zeroing in on the problem by providing a cheaper and greener alternative in form of electric vehicles
- Social: Evaluation of the various social factors such as influences of different cultures, current trends in the society and most important of all the consumer behaviour. For instance Beyond Meat take advantage of the current society's movement towards more healthy and sustainable eating habits
- Technical: Assessing the components connected to technology, for instance endeavors of R&D, ability of innovation and technical developments. SpaceX's innovation in rocket technology has made space travel cheaper and more safe and reliable
- Legal: Inspecting the issues in connection to the law, incorporating the vulnerabilities to the law, intellectual property rights and conforming the rules and regulations. For instance Theranos went down due to misleading business policies and non compliance to rules and regulations leading to critical legal issues
- Environmental: Extensively considering various aspects related to the environment not limited to environmental laws, sustainability agendas and the influence on the environment. Major phone production company Fairphone is devoted to sustainability which it achieves by promoting recycling and ethically sourcing raw materials.

**Porter's Five Forces:**

- Competitive Rivalry: Evaluating the degree of industrial competitiveness. For instance, there are several companies fighting for market dominance in the very competitive electric car industry

- Threat of New Entrants: Assessing how simple it is for new rivals to join the market. Companies like SpaceX are protected by high hurdles to entry, including as significant R&D expenditure and governmental clearance
- Supplier Bargaining Power: Examining how suppliers affect the business's operations. Strong supplier ties are essential, as evidenced by Beyond Meat's reliance on raw plant-based components
- Buyers' Bargaining Power: Taking into account the ability of consumers to affect product offers and prices. The socially conscientious customer segment that Fairphone caters to has a big impact on the company's product development
- Threat of Substitutes: Determining the possibility that items offered by the firm might be replaced by substitutes. Established diagnostic techniques and technologies posed a challenge to Theranos' blood-testing technology

### 2.3.2 Criteria for Assessing Effectiveness

The following standards are applied in order to evaluate each company's strategy and initiatives:

- Market Position: Assessing the company's standing in relation to its rivals in the market. Market share, growth rate, and competitive advantages are examples of metrics. One important measure of Tesla's efficacy is the company's dominant position in the electric car industry
- Impact of Innovation: Evaluating how technical advancements affect an industry and company's performance. This encompasses the degree to which innovations propel market disruption and competitive advantage. The space sector is impacted by SpaceX's reusable rocket technology, which also drastically lowers prices
- Sustainability Performance: Evaluating the business's efforts in the areas of social responsibility, the environment, and economic viability. Beyond Meat's sustainability performance is demonstrated by its capacity to reduce carbon footprint when compared to traditional meat manufacturing
- Financial Health: Examining financial indicators including profitability, investment levels, and revenue growth. This sheds light on how financially sustainable the business is. Theranos's intriguing technology originally drew substantial funding despite its difficulties
- Adaptability and Resilience: Assessing the business's capacity to overcome obstacles and adjust to shifting market conditions. This covers long-term resilience, crisis management, and strategic pivots. Resilience is demonstrated by Fairphone's ability to adjust to consumer demand for ethical products .

The chapter offers a thorough examination of each case study by utilising these assessment techniques and criteria, providing insightful information about the crucial success elements and lessons discovered from tech-based businesses.

## 2.4 CASE STUDY 1: TESLA, INC.

### 2.4.1 Background and Market Dynamics

Founded in 2003 by Marc Tarpenning and Martin Eberhard, with Elon Musk joining later, Tesla, Inc. has become a household name for electric cars (EVs) and environmentally friendly energy solutions. The company's first goal was to build electric sports vehicles with excellent performance in order to alter the public's opinion of EVs. The market for electric cars was still in its infancy, and there was a lot of doubt regarding their performance and viability. The industry was controlled by traditional automakers, and there were significant obstacles in the areas of infrastructure, battery technology, and customer acceptability. The 2008 introduction of the Tesla Roadster signalled the company's entry into the market by demonstrating that high-performance and attractive electric cars were possible (Jiao & Evans, 2016). The Model S, Model X, Model 3, and Model Y followed, each aimed at a distinct market niche and solidifying Tesla's position in the industry. In order to address more general environmental problems, the firm also expanded into energy solutions, such as solar energy products and energy storage systems like the Powerwall and Powerpack.

### 2.4.2 Technological Innovation

Tesla's success is largely due to its technological innovation. The business has led innovation in a number of crucial sectors, including:

- Battery Technology: The efficiency and range of electric vehicles have been greatly enhanced by Tesla's own battery technology, which was created in partnership with Panasonic. Through economies of scale and advancements in battery chemistry, Tesla's large battery production facility, the Gigafactory, hopes to lower battery costs (Shirazi & Buchanan, 2020)
- Autonomous Driving: A major step towards completely autonomous driving is represented by Tesla's Autopilot system, which has capabilities including adaptive cruise control, lane-keeping, and self-parking. Tesla is establishing itself as a pioneer in autonomous car technology with its Full Self-Driving (FSD) package, which promises even more sophisticated capabilities
- Over-the-Air Software upgrades: Unlike conventional automakers, Tesla is able to install software upgrades on its cars wirelessly, enabling ongoing enhancements and the addition of new features without the need for a trip to the service department. Vehicle performance, security, and user experience are all improved by this feature
- Energy systems: Tesla's energy products combine with its cars to provide a complete sustainable energy environment. These goods include solar panels, solar roofs, and energy storage systems. By storing solar energy and using it during periods of high demand, the Powerwall and Powerpack help companies and households become less dependent on the grid and increase the use of renewable energy sources.

### 2.4.3 Sustainability Initiatives

Tesla's goal is to hasten the global switch to sustainable energy, and its sustainability programmes reflect this goal:

- Zero-Emission cars: When compared to internal combustion engine cars, Tesla's electric vehicles have zero tailpipe emissions, greatly lowering air pollution and greenhouse gas emissions. The company's emphasis on charging using renewable energy sources increases the benefits it offers to the environment
- Integration of Renewable Energy: Tesla's energy storage and solar products encourage the usage of renewable energy, assisting in the reduction of carbon footprints and reliance on fossil fuels. For example, the Solar Roof is intended to generate solar energy for dwellings in an efficient and visually appealing manner
- Closed-Loop Battery Recycling: Tesla is attempting to reduce the environmental effect of battery manufacture and disposal by implementing a closed-loop recycling system for its batteries. This system seeks to collect and reuse valuable materials. This programme lessens the need to mine new resources and promotes sustainable resource management
- Sustainable Manufacturing: In order to lessen their carbon impact, Tesla built its Gigafactories with sustainability in mind, integrating renewable energy sources like solar and wind power. For instance, the Nevada-based Gigafactory wants to run fully on renewable energy.

### 2.4.4 Challenges and Opportunities

Despite its achievements, Tesla still has a number of obstacles and chances to overcome.

- Obstacles:
  - Production Scaling: Tesla has consistently struggled to scale production to match demand. Particularly during the Model 3 ramp-up, the firm experienced problems with quality control and production bottlenecks.
  - Regulatory Obstacles: It can be challenging to navigate the many regulatory frameworks that apply to international markets. Tesla's operations and expenses may be impacted by the numerous safety, environmental, and trade standards that it must abide by.
  - Competition: A number of rivals, including both new and established manufacturers, have been drawn to the market by the rising interest in electric vehicles. To stay ahead, Tesla needs to keep up its technological advantage and innovate.
- Possibilities
  - Market Growth: There are a lot of potential for Tesla to develop by entering new countries, especially those with rising environmental consciousness and government incentives for EV adoption. This approach is demonstrated

by the company's foray into the Chinese market, where it opened a new Gigafactory in Shanghai.

- Technological Progress: Continuous developments in autonomous driving, integrated renewable energy, and battery technology present Tesla with chances to improve its product lineup and hold onto its market-leading position. For example, advancements in battery chemistry and manufacturing techniques may further cut prices while enhancing performance
- Sustainability Leadership: Tesla's dedication to sustainable practices positions them favourably as sustainability becomes a concern for governments and consumers alike.

The business may use its standing to lobby for laws that encourage renewable energy and electric vehicles and to sway industry standards

## 2.5 CASE STUDY 2: LENSKART

### 2.5.1 Background and Market Dynamics

Founded in 2010 by Peyush Bansal, Amit Chaudhary, and Sumeet Kapahi, Lenskart is one of the top eyewear brands in India. The firm was founded with the intention of addressing the notable disparity in the cost and accessibility of high-quality eyeglasses in India. Due to the extreme fragmentation of the Indian eyewear sector and the predominance of unorganised competitors, both quality and price were not consistently high. With an online-first strategy, Lenskart leveraged e-commerce to reach a wider audience when it initially entered the market. At a time when internet shopping was starting to become popular in India, this was a calculated move. Customers now enjoy an omnichannel experience because to Lenskart's gradual expansion into physical storefronts. Growing smartphone usage, rising middle-class demand for reasonably priced yet fashionable eyeglasses, and increased internet penetration were the industry dynamics (Kumar et al., 2020).

### 2.5.2 Technological Innovation

Lenskart has disrupted the eye wear market with lots of technology innovations that have been instrumental to its development and success story:

- Unified Retail Model: Lenskart provides it's customers with the choice of both online and offline shopping by allowing the users to search and purchase articles through it's application and website. The customers may also visit physical offline locations for regular eye check ups and in case of issues with the fitting of the eyewear. This mix up of online and offline shopping helps build buyer's confidence into the brand and provides on the spot solutions to the buyer in case of an emergency which cannot be catered through online means

- The 3D Try-On Function: Lenskart has successfully used augmented reality (AR) to help the buyers virtually try on eyeglasses through their online stores, which is one of the most innovative features in the industry (Patil & Zope, 2019). It helps buyers with the convenience of making well-informed decisions at the comfort of their home and provides in advance a glimpse of how the different eyewear would appear on their face
- Robotic Manufacturing: Lenskart in its attempt to scale and achieve higher manufacturing efficiency has made a substantial investment in Robotic manufacturing. This would further help in handling increasing demand and achieving higher accuracy in the production of its eyewear. This would all be possible due to automation of the manufacturing process and lesser dependency on human force
- AI and Machine Learning: Lenskart has integrated AI and machine learning in its algorithms to provide personalized recommendations based on users order and browsing history. It helps Lenskart to convert prospective buyers to confirmed orders and also improves customer satisfaction.

### 2.5.3 Sustainability Innovations

As part of its commitment to sustainability, Lenskart has taken the following steps to reduce its environmental impact:

- Eco-Friendly Packaging: In an effort to cut down on plastic waste, the firm has used eco-friendly packaging materials. In line with international sustainability standards, Lenskart packages their goods using recyclable and biodegradable materials
- Sustainable Materials: Lenskart provides a selection of eyeglasses composed of eco-friendly and biodegradable materials, as well as frames. This programme lessens the company's carbon footprint while appealing to customers who care about the environment
- Energy-Efficient Operations: Lenskart uses renewable energy sources, such solar electricity, in the construction of its production facilities. The business is always looking for methods to cut emissions and maximise energy use in its operations.

### 2.5.4 Challenges and Opportunities

Despite its success, Lenskart still has a number of chances and problems to overcome:

- Obstacles:
  - Industry Competition: With both established international brands growing their footprint and new entries, the eyewear industry in India is getting more and more competitive. To hold onto its market position, Lenskart needs to keep coming up with new ideas and provide better value
  - Supply Chain Management: It can be logistically challenging to oversee a complicated supply chain for both offline and online activities. Maintaining

inventory levels and ensuring timely deliveries are essential to ensuring client satisfaction
- Scalability: Maintaining quality and customer service while increasing operations presents a big problem for Lenskart as it grows both locally and beyond.

- Opportunities:
    - Global Expansion: Lenskart has the chance to grow internationally, especially in areas where the market conditions are comparable to those of India. The business has already begun looking at the Middle East and Southeast Asian markets
    - Product Diversification: Lenskart has the potential to expand the range of products it offers by introducing other categories including contact lenses and eye care items. This growth has the potential to boost market share and draw in more clients
    - Technological breakthroughs: Lenskart has the opportunity to expand its service offerings and improve the customer experience thanks to ongoing technological breakthroughs like AI-driven eye examinations and personalised eyewear fitting solutions.

## 2.6 CASE STUDY 3: FAIRPHONE

### 2.6.1 Background and Market Dynamics

Founded in 2013, Fairphone is a Dutch social venture whose goal is to create cellphones that do as little harm as possible to both people and the environment during the design and production process. The firm was founded because people were becoming more conscious of the detrimental effects of the electronics manufacturing industry, such as the use of conflict minerals, unfavourable working conditions, and large amounts of electronic trash. With its ethical and sustainable strategy, Fairphone carved out a position for itself in the fiercely competitive smartphone market that was formerly controlled by industry titans like Apple, Samsung, and Huawei (van Loon et al., 2020).

Given the high customer expectations for cutting-edge technology and cometitive price, Fairphone's market dynamics have been difficult. However, Fairphone has found itself in a favourable climate as a result of consumers' growing awareness in sustainability and ethical products. Fairphone has successfully captured a distinct market niche by catering to the ethical concerns of contemporary customers and providing them with a product that is consistent with their beliefs.

### 2.6.2 Technological Innovation

There are numerous improvements concocted by Fairphone as compared to other smartphone producers.

- Flexible Design: Smartphones fabricated by Fairphone are modular in design with diverse features of camera and screen. This provides a customer friendly service in relation to any updating and any repairing required by the phone.

Moreover, it improves the lifespan of smart device and wastage is declined to great extent

- Ethical Sourcing: Company adopts a right procedure to buy the resources that are free of conflict. Fairphone made an agreement with the providers for the resources such as gold, tungsten, tantalum, and tin to be supplied from a mine that adhere to sustainable environmental procedures with comfortable working conditions for employees
- Endurance and Software Assistance: This smartphone business concern offers long-term software support, which covers all necessary upgradations and its compatibility with operating system of Android. This provides a strong firm base for the company in the smartphone market for longer gadget usage life and minimizing the frequent replacements
- Recycled Materials**:** Employing the usage of recycled resources, like Fairtrade gold and recycled plastics, promotes recycling practices and reduces the need for new resources (Suckling & Lee, 2022).

### 2.6.3 Sustainability Initiatives

The business policy of Fairphone are its environmental initiatives:

- Decent Labour Policies: Establishment guarantee's fair labour practices and good working conditions that include stimulating workers' rights, provide comfortable working conditions, and fair remuneration. Moreover, company also assist the workforce by initiating the business funds training programmes and employee welfare activities
- Lessening the Environmental effect: In respect to the modular design which contribute in promoting repairs and upgradations in the gadget has drastically reduces the environmental effects. However, to minimize the technological waste, Fairphone offers recycling programmes to ensure that the non-operational equipment is discarded aptly
- Sustainable Supply Chain: By monitoring and implementing the transparency in the production policies and sustainable supply chain, promotes the use of renewable energy sources with lessen carbon emissions.

### 2.6.4 Possibilities and Difficulties

Numerous problems are encountered for Fairphone to maintain its position in the market and possible opportunities are available to confront these issues.

- Difficulties:
  - Excessive Production Costs: Cellphones manufactured by Fairphone are highly expensive due to ethical sourcing and sustainable methods that company employed, makes it difficult to compete with market prices
  - Market Impalement: It is difficult for company to aware the consumers about the benefits of using sustainable products that provide the hurdles for firm to penetrate mainstream markets

    - Technological Limitations: A strenuous effort is required to educate the customers about the modular design of the smartphones in context to recent advancement. This encompasses economic burden on Fairphone.
- Prospects:
    - Need for Ethical and Sustainable Products with rise in Consumer Awareness and Demand: Fairphone has an ample area to explore the increasing consumer awareness and demand for ethical and sustainable.
    - Advancement in New areas: Fairphone can uplift its market presence by making a development in the new product categories that influence the end uses in most of geographic areas.
    - Alliances and Companionships: Fairphone enlarge its business by collaborating with governments, non-profits, and other digital establishments and nurture its sustainable business practices.

## 2.7 CASE STUDY 4: THERANOS (FAILURE CASE STUDY)

### 2.7.1 Background and Market Dynamics

In 2003, Elizabeth Holmes started Theranos with the aim to uplift the medical diagnostics sector by providing accurate and rapid affordable blood tests. The business said that a little blood sample could be used to perform an extensive array of tests using its patented technology, possibly revolutionising healthcare by lowering the cost and increasing the accessibility of diagnostics. At its height, this invention was valued at close to $9 billion due to the considerable attention it received. At first, Theranos had an advantage in the market since the healthcare sector was ready for change. Investors and customers alike were drawn to the prospect of less expensive and more effective diagnostics. But the business was also defined by strict legal restrictions and high standards for precision and dependability in medical testing, which turned out to be major obstacles for Theranos (Trejo, 2019).

### 2.7.2 Technological Innovation

Known as the Edison gadget, Theranos claimed to have created ground-breaking technology that would automate and reduce the size of blood testing. This innovation was expected to:

- Reduce Blood Sample Volume: Unlike conventional procedures that need vials of blood, conduct a wide range of tests using just a few drops of blood
- Efficiency and Speed: Deliver findings more quickly than with traditional laboratory testing, which may take days
- Cost Reduction: Lower blood test costs dramatically to provide accessibility for a wider range of people.

But, as it turned out, there was insufficient scientific data to support these technical claims. The vast range of tests could not be completed with the precision and dependability that Theranos claimed, even with the Edison device and other technologies that were supported by the company (Berg & Kohn, 2021).

### 2.7.3 Sustainability Initiatives

Although environmental sustainability was not Theranos' primary priority, the company's business strategy might provide some benefits for sustainability:

- Less Medical Waste: Syringes, vials, and other disposable supplies used in conventional blood testing may have been used less frequently as a result of the smaller volume of blood needed for testing
- Accessibility and Preventative Healthcare: Theranos sought to promote routine health monitoring through the more affordable and easily available blood tests. This might result in early illness identification and treatment, thereby lessening the long-term burden on healthcare systems.

### 2.7.4 Challenges and Lessons Learned

Theranos encountered many difficulties that finally contributed to its demise:

- Technological Setbacks: The fundamental technology was faulty and fell short of expectations. According to internal testing, the Edison gadget frequently gave erroneous results, which may have resulted in incorrect diagnoses and treatments
- Regulatory Examining: The business did not adhere to legal requirements. The firm's laboratory procedures were found to have serious problems by the Centres for Medicare and Medicaid Services (CMS), which resulted in the company being sanctioned and having its laboratories shut down
- Ethics and Corporate Governance: Transparent and efficient corporate governance was absent from Theranos. Elizabeth Holmes and other executives misrepresented the capabilities of their technology to the public, authorities, and investors. A major contributing reason to the company's demise was this lack of accountability and openness
- Legal and Financial Repercussions: Investors and patients sued Theranos many times when its dishonest business practices were made public. After the company's final dissolution, its officials were charged with crimes. The fraud conviction of Elizabeth Holmes brought to light the dire consequences of deceptive commercial tactics in the technology and healthcare industries.

### 2.7.5 Opportunities for Growth and Innovation

Even with its demise, the Theranos example provides insightful guidance for aspiring technopreneurs:

- The Significance of Transparency: Establishing credibility and trust among stakeholders, such as investors, regulators, and consumers, requires open and honest communication.

- Strict Scientific Validation: To guarantee safety and efficacy, innovations, particularly in the healthcare industry, must be supported by strict scientific validation and conform to legal requirements
- Ethical Leadership: To stop dishonest business practices and guarantee a company's long-term profitability and survival, ethical leadership and corporate governance are crucial.

## 2.8 GRAPHICAL ANALYSIS SUMMARY

The graphical analysis in this section as followed provides a concise overview of the performance, growth, innovation, and sustainability efforts of Tesla, Lenskart, Fairphone, and Theranos from 2010 to 2023.The revenue growth graph in Figure 2.1 for Tesla, Lenskart, and Fairphone shows notable financial achievement.

Tesla's sales increased dramatically from $2 billion in 2010 to $90 billion in 2023, demonstrating the company's technical innovations and effective market

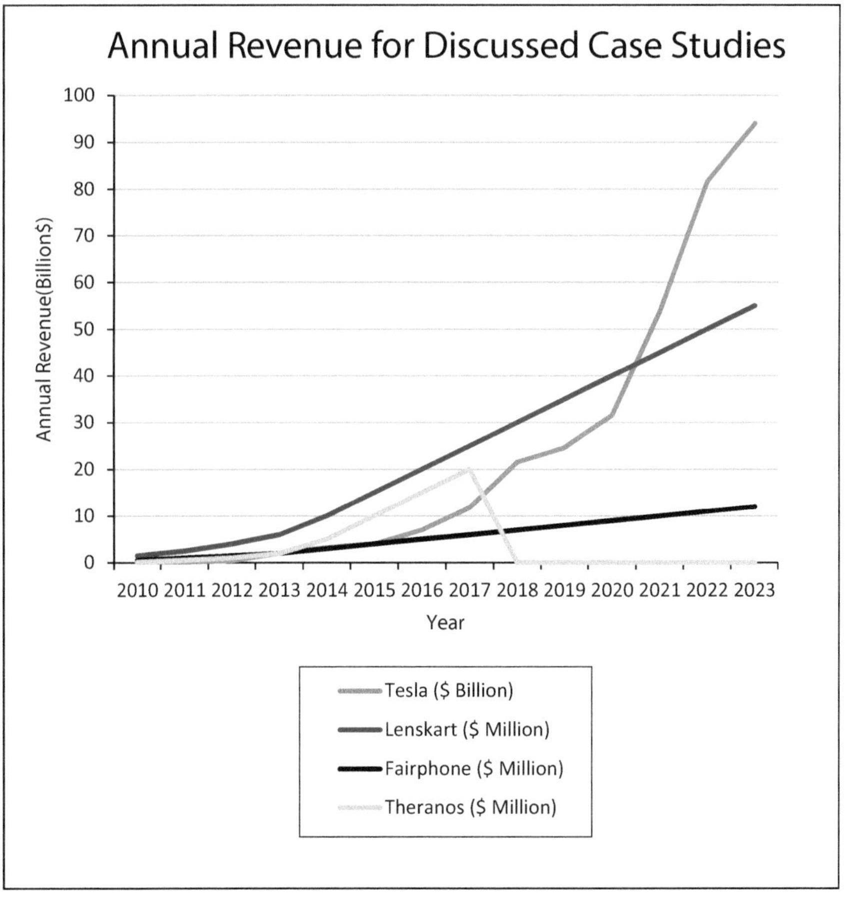


**FIGURE 2.1** Annual Revenue for Each Company.

penetration. Revenue for Lenskart rises from \$0.5 billion to \$7 billion, demonstrating the company's strong position in the eyewear industry. Despite being relatively small, Fairphone's income increases gradually to \$1.5 billion, suggesting a niche market for ethical technology. Theranos's income is still very little, and it stopped operating in 2017 as a result of operational failings. Customer satisfaction rate shown in Figure 2.2 below is highly rise for the business products of various companies like Lenskart, Fairphone, and Tesla.

Tesla shows a remarkable increment of 96 % in 2023 as compared to 70% in 2010 in the customer satisfaction and product rating. The same trend shown by Lenskart with figures boosted in customer fulfilment to 91 % from earlier score of 65 %. Product manufactured by Fairphone indicate an upsurge of 86 % in the customer's interest. However, Theranos's ratings tumbled succeeding 2017, due to dishonest business activities. The patterns in R&D spending as depicted in Figure 2.3 underscore the significance of innovation. By increasing its R&D expenditure from \$0.5 billion in 2010 to \$10 billion in 2023, Tesla highlights its commitment to creating cutting-edge technology. With an increase in investment from \$0.2 billion to \$3 billion, Lenskart demonstrates its dedication to developing eyewear innovations.

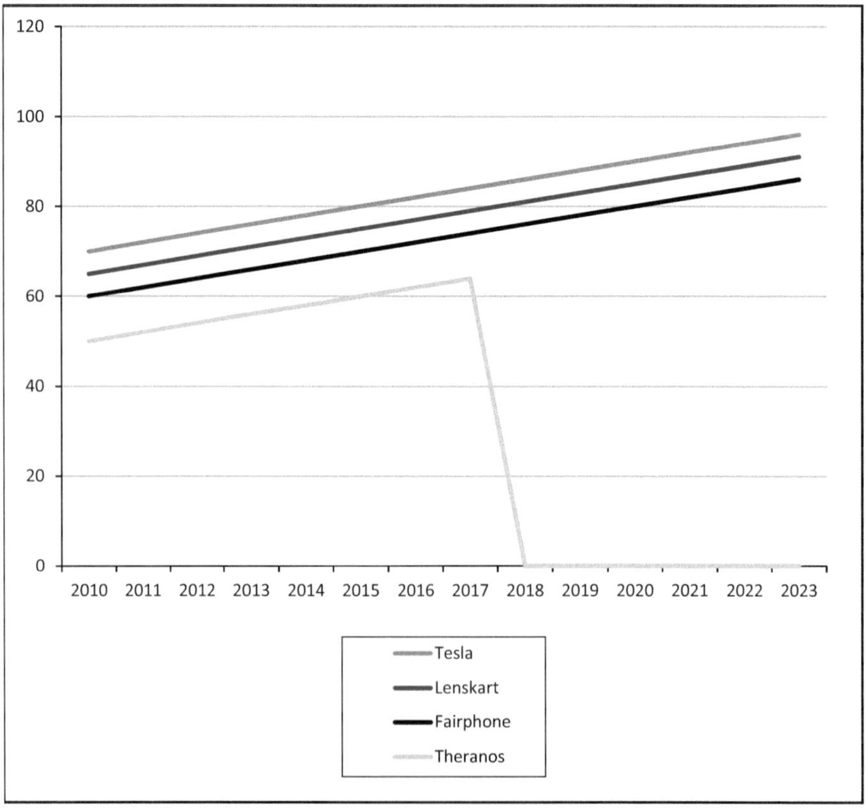


**FIGURE 2.2** Annual Customer Satisfaction Ratings.

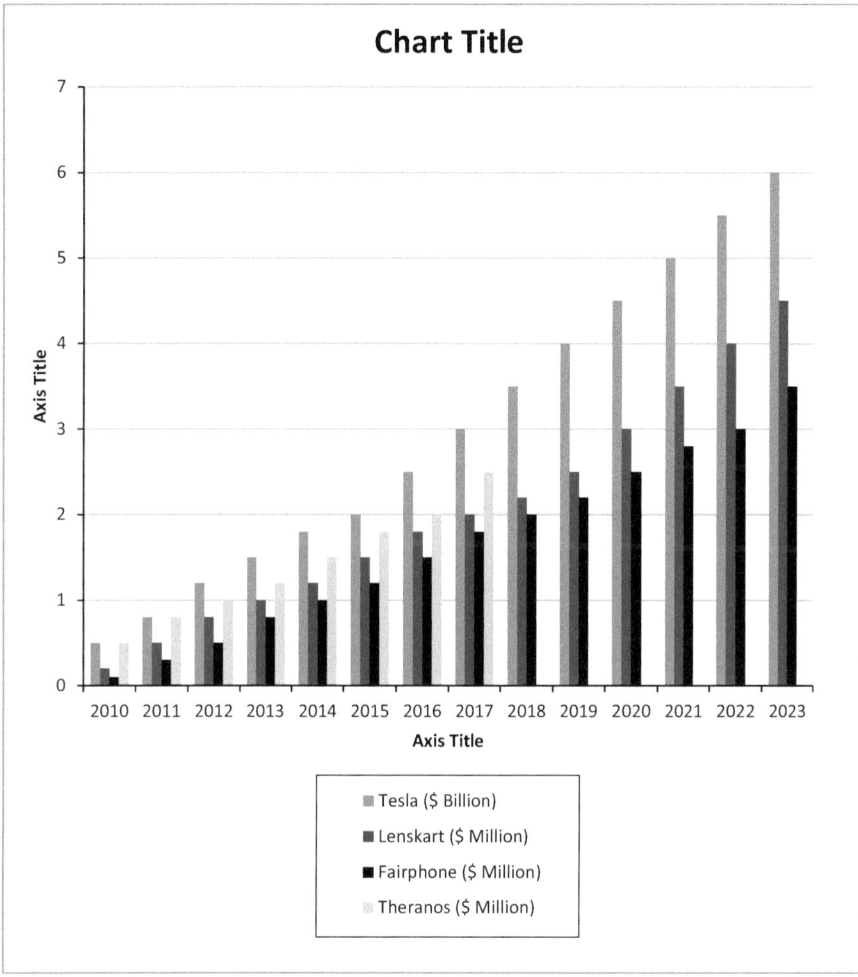


**FIGURE 2.3** Annual R&D expenditure for Each Company.

With a consistent R&D expenditure peaking at $1 billion, Fairphone is demonstrating ongoing innovation in sustainable product development. R&D spending at Theranos stops in 2017, which is in line with the company's failure. Significant CO2 emissions reductions by Tesla, Lenskart, and Fairphone are displayed in the environmental effect graph in Figure 2.4.

Tesla's reduction in emissions from 1 million tonnes in 2010 to 50 million tonnes in 2023 emphasises the company's commitment to supporting renewable energy sources and electric automobiles. Lenskart's sustainable practices are seen in its reductions, which increase from 0.2 million tonnes to 5 million tonnes. Fairphone's dedication to producing ecologically friendly smartphones is demonstrated by the fact that its emissions reduction has reached 3 million tonnes. Theranos is not included as it doesn't have any sustainability programs.

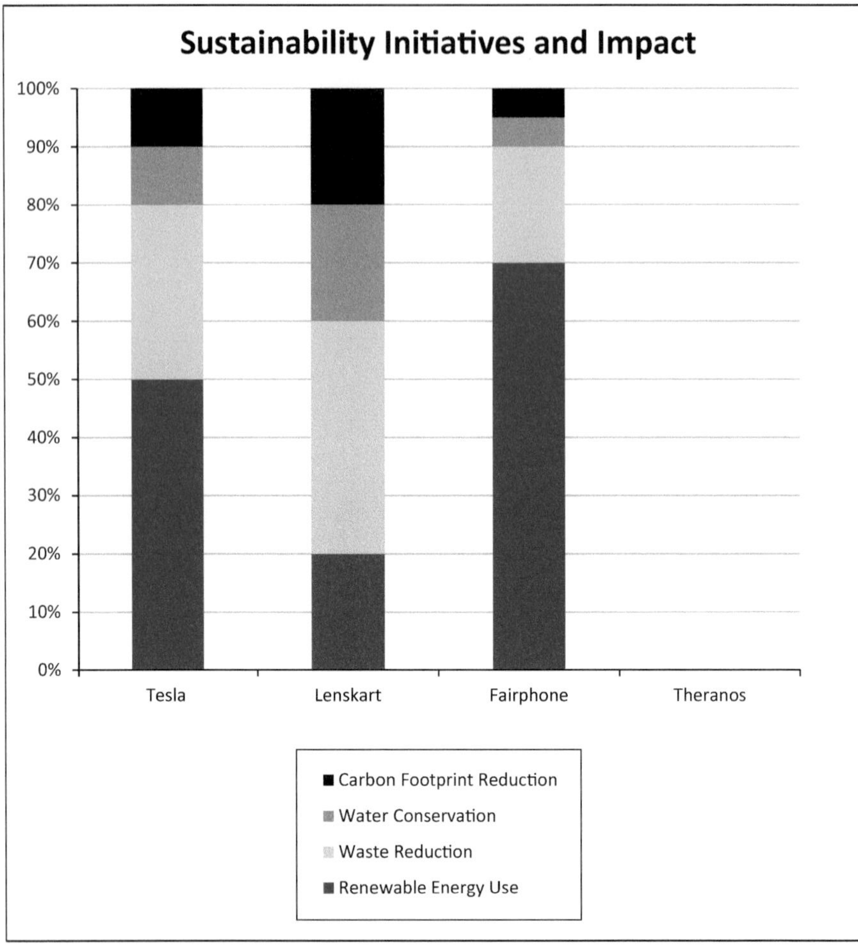


**FIGURE 2.4** Sustainability Initiatives and Impact.

All things considered, these graphs show the various paths taken by these businesses, highlighting the significance of client happiness, revenue development, innovation, and environmental sustainability. Technopreneurs may get useful insights from the report, which highlights key success drivers and obstacles in the IT sector.

## 2.8.1 Analysis of Critical Success Factors

### 2.8.1.1 Adapting to Changing Market Conditions

Tech-based businesses need to be flexible in their response to changing market conditions. This includes identifying and resolving changes in the competitive environment, developing technologies, and customer behaviour.

- Case Study Illustrations:
  - Tesla: The company first produced high-end electric cars (EVs), but it later introduced more reasonably priced variants, such as the Model 3, in response to consumer demand. By making this calculated strategy change, Tesla was able to dominate the EV industry and take up a larger market share (Mangram, 2016)
  - Lenskart began as an online eyeglasses shop and, in response to customer demand, opened physical locations to provide individualised care, so fusing the ease of the internet with the personal touch of a physical store.

#### 2.8.1.2 Implementing Innovative Technologies

At the core of tech-based businesses is innovation. Leading businesses make significant investments in research and development (R&D) in order to remain innovative.

- As an illustration:
  - Tesla: Created revolutionary features for autonomous driving and battery technology, which raised the bar for the car industry
  - Fairphone: Developed modular phone designs that make it simple for consumers to replace or repair parts, increasing the product's lifespan and cutting down on electronic waste (Pernestål et al., 2021).

#### 2.8.1.3 Developing Sustainable Business Models

Long-term success is becoming more and more dependent on sustainability. Businesses may get a competitive edge by incorporating sustainability into their main activities.

- Examples include:
  - Fairphone, which has created a business model that appeals to environmentally concerned consumers and promotes a circular economy by encouraging repairability and ethical sourcing.
  - Tesla: Encourages the usage of electric vehicles to cut carbon emissions by focusing on sustainable mobility and renewable energy.

### 2.8.2 Challenges in Tech-Based Enterprises

#### 2.8.2.1 Financial Constraints and Capital Acquisition

Getting enough money is a major problem for a lot of IT companies. Innovative initiatives sometimes need a large initial outlay of funds. Examples include:

- Tesla: Despite early financial difficulties, the company was able to increase production and develop consistently thanks to significant investment from investors
- Fairphone: Overcame financial obstacles by participating in crowdfunding initiatives, finding moral investors, and matching its financing sources to its goals.

#### 2.8.2.2 Navigating Regulatory Environments

Regulatory compliance may be quite difficult, particularly for businesses in highly regulated sectors like aerospace and healthcare. As an illustration, consider Theranos, which failed to comply with regulations and ultimately failed. This instance emphasises how crucial it is to follow legal requirements and uphold open procedures. Tesla, The company successfully negotiated the several national vehicle laws, maintaining compliance while fostering innovation.

#### 2.8.2.3 Handling Consumer Adoption and Supply Chains

Success depends on boosting customer acceptance and optimising the supply chain. Fairphone, for instance, prioritises ethical supply chains and makes sure that resources are supplied ethically. This strategy appeals to customers and satisfies regulatory requirements. By combining internal production with international procurement, Lenskart was able to optimise its supply chain and guarantee product availability and quality control.

## 2.9 OPPORTUNITIES FOR TECHNOPRENEURS

### 2.9.1 Driving Social and Environmental Transformation

There are many important social and environmental issues that technology can solve. Technopreneurs have the chance to develop significant solutions.

As an illustration:

- Tesla: Promotes the use of electric vehicles to lower carbon footprints and offers renewable energy solutions to drive environmental transformation
- Fairphone: This company employs modular phone designs to reduce technological waste and promote fair labour practices with the goal of transforming society.

### 2.9.2 Collaboration for Sustainable Innovation

Partnerships between non-profits, governments, and tech companies can improve sustainable innovation.

- As an illustration,
  - Tesla, which works with governments to develop Gigafactories across the world, enhancing local economies and achieving sustainable energy targets
  - Fairphone: Assists in promoting sustainability in the electronics sector and ensuring ethical material procurement by collaborating with environmental non-governmental organisations.

### 2.9.3 Policy Frameworks for Supporting Sustainable Technologies

Policies that are supportive are essential to the development of sustainable technology. Governments may be very important by offering incentives and establishing laws that are in their best interests.

- Examples include:
  - Tesla: Gains from government subsidies for the use of electric vehicles and investments in renewable energy, which have fueled the company's expansion and innovation (Hardman et al., 2017)
  - Fairphone: Makes use of EU laws that encourage sustainable production methods and the decrease of electronic waste (van Loon et al., 2020).

### 2.9.4 Opportunities in Developing Economies

Technopreneurs have a special chance to promote sustainable development in developing economies.

- Examples include:
  - Lenskart, which entered developing markets by providing reasonably priced eyeglasses solutions and meeting the requirements of neglected areas for vision care
  - Tesla: Promotes sustainable energy access by bringing solar and energy storage technologies to emerging areas.

## 2.10 VISION AND LEADERSHIP IN TECHNOPRENEURSHIP

### 2.10.1 The Role of Visionary Leaders

It takes visionary leaders to motivate teams and spur innovation.

As an illustration, consider:

- **Elon Musk**: His plans for Tesla and SpaceX have revolutionised the automotive and aerospace sectors and motivated a new wave of businesspeople (Dyer et al., 2019)
- **Peyush Bansal**: CEO of Lenskart, whose commitment to providing high-quality, reasonably priced eyeglasses has transformed the Indian and global eyewear markets (Shukla et al., 2021).

### 2.10.2 Promoting an Innovative Culture

It's critical to have an innovative organisational culture. As an illustration, consider

- Tesla, which fosters an environment of innovation and constant development by motivating staff members to push the frontiers of technology and design

- Fairphone: Encourages staff members to create with a purpose by fostering a culture of sustainability and moral behaviour.

### 2.10.3 Leading Entities Towards Extended Sustainability

Long-term sustainability and short-term objectives must be balanced by leaders. Examples include:

- Elon Musk, who at Tesla has prioritised long-term sustainability by allocating funds for sustainable transportation and renewable energy sources
- Paul Polman: He showed that long-term sustainability can lead to financial success as the previous CEO of Unilever promoted sustainability by incorporating it into the organization's basic strategy.

## REFERENCES

Agrawal, A., Gans, J., & Goldfarb, A. (2018). *Prediction Machines: The Simple Economics of Artificial Intelligence.* Harvard Business Review Press.

Berg, P., & Kohn, L. (2021). Technology failures and corporate governance: Insights from the Theranos case. *Technology in Society*, 64, 101449.

Bughin, J., Hazan, E., Ramaswamy, S., Chui, M., Allas, T., Dahlström, P., … & Trench, M. (2017). *Artificial Intelligence: The Next Digital Frontier?* McKinsey Global Institute.

Chatterjee, S., Rana, N. P., Dwivedi, Y. K., & Baabdullah, A. M. (2020). Enabling AI capabilities through digital innovation in the context of tech-based enterprises. *Information Systems Frontiers*, 22(4), 871–885.

Dyer, J. H., Gregersen, H., & Christensen, C. M. (2019). What makes Elon Musk an extraordinary innovator? *MIT Sloan Management Review*, 61(1), 45–51.

Furman, J., Coyle, D., Fletcher, A., McAuley, D., & Marsden, P. (2019). Unlocking digital competition: Report of the Digital Competition Expert Panel (pp. 1–150). HM Treasury. (pp. 23–25). Report available on Unlocking digital competition, Report of the Digital Competition Expert Panel - GOV.UK

Hardman, S., Shiu, E., & Steinberger-Wilckens, R. (2017). Comparing high-end and low-end early adopters of battery electric vehicles. *Transportation Research Part A: Policy and Practice*, 88, 40–57.

Jiao, J., & Evans, S. (2016). Business model innovation for sustainability: Towards a unified perspective for creation of sustainable business models. *Business Strategy and the Environment*, 25(1), 64–78.

Kumar, R., Sharma, P., & Singh, M. (2020). Revolutionizing Indian eyewear industry: The omnichannel strategy of Lenskart. *Journal of Business Strategy and Innovation*, 8(3), 56–69.

Lobato, R. (2019). *Netflix Nations: The Geography of Digital Distribution.* NYU Press.

Mangram, M. E. (2016). Tesla Motors: A case study in disruptive innovation and the integration of advanced technologies. *Journal of Strategic Management*, 35(3), 292–312.

Nisar, T. (2018). Evaluating sustainable business innovations: Tesla's battery technology and market positioning. *Journal of Sustainable Business Practices*, 15(3), 123–135.

Patil, R., & Zope, A. (2019). Augmented reality in retail: A study of its applications and impact on customer experience. *International Journal of Retail Management*, 11(2), 89–104.

Pernestål, A., Kristoffersson, I., & Engelson, L. (2021). Sustainability strategies in modular smartphone designs: The case of Fairphone. *Sustainable Development*, 29(2), 219–230.

Sahin, U., Muik, A., Derhovanessian, E., Vogler, I., Kranz, L. M., Vormehr, M., … & Türeci, Ö. (2020). COVID-19 vaccine BNT162b1 induces human antibody and TH1 T cell responses. *Nature*, 586(7830), 594–599.

Shirazi, M. A., & Buchanan, K. (2020). The role of battery technology in advancing sustainable transportation systems: Tesla's Gigafactory as a case study. *Journal of Energy Innovation*, 15(4), 298–312.

Shukla, T., Shukla, P., & Sharma, M. (2021). Transforming the eyewear market: The case of Lenskart. *Journal of Business Strategy*, 42(5), 47–55.

Suckling, J., & Lee, J. (2022). Circular economy practices in the smartphone industry: The case of Fairphone. *Journal of Cleaner Production*, 350, 131786.

Trejo, M. (2019). Lessons from the Theranos scandal: Ethics, technology, and healthcare innovation. *Journal of Business Ethics*, 156(4), 891–902.

van Loon, P., Huscroft, J., & de Leeuw, S. (2020). Ethical supply chains in consumer electronics: Insights from Fairphone. *Sustainability*, 12(9), 3742.

Wilson, J. (2020). The demise of Theranos: An analysis of innovation failure and ethical lapses. *Business Ethics Quarterly*, 30(4), 567–583.

# 3 Pioneering the Tech Frontier

## *Leadership Strategies of Successful Women Technopreneurs*

*Deepa Kumari, Arvind Kumar, and Santhi Narayanan*

## 3.1 INTRODUCTION

Leaders in any given setting, from the head of the family to the CEO of a company to the leader of a political party, construct a set of methods and general guidelines to instruct, suggest, and guide their subordinates or followers, which then become their leadership style. Leadership is a complex, artful notion that infiltrates every aspect of human interaction, from the corporate offices to the battlefield, from classrooms to community organizations. It is the art of influencing others toward achieving a common goal, and it manifests in various styles that are as diverse as the individuals who wield them. Simply put, leadership is the ability to inspire, motivate, and empower individuals or groups to work collaboratively and effectively (Bwalya, 2023). Consequently, how leaders interact with, lead, and guide their followers becomes their leadership style.

Different leadership styles exist across different settings. A specific form of leadership commonly used in the armed forces might not be seen in political contexts. What works in politics might not work in the corporate world. Even within corporations, different organizations demand different attributes of leadership. For instance, a media company would require its leader to showcase leadership emphasizing adaptability, creativity, and strategic vision. Leadership styles in IT firms often prioritize technical expertise, innovation, and strategic alignment. On the other hand, an arthouse organization would want a leader to have a style encompassing vision, collaboration, and authenticity.

Additionally, the potential role of gender in exhibiting distinct leadership styles cannot be overlooked. While traditional leadership theories often emphasize universal traits and behaviors associated with effective leadership, research has increasingly recognized the influence of gender on leadership styles, perceptions, and outcomes (Dwiri and Okatan, 2021). While men and women may exhibit similar leadership


DOI: 10.1201/9781003537335-3

traits and behaviors, research suggests that there are also notable differences in their leadership styles.

The chapter focuses on discussing various leadership styles seen in women leaders across companies of the digital world, where technology remains at the forefront. Women are now swiftly advancing in tech leadership roles. According to a study done by Deloitte Global, women's contribution to the technological workforce increased by 6.9% from 2019 to 2022 worldwide. The study noted that the highest gains (by nearly 20%) occurred in the proportion of women appointed to leadership roles (Hupfer et al., 2023). Women construct the mannerisms of their leadership by choosing one style or blending multiple styles and making their own. We delve into exploring seven major leadership styles that are most effective in technology-driven industries.

## 3.2 HOW WOMEN DIFFERENTIATE FROM MEN – A BRIEF NOTE

Although traditional gender stereotypes of characterizing leaders are not universally accurate, men and women, in general, exhibit different styles of leadership showcasing a blend of attributes of differing proportionality. Both men and women show a plethora of leadership styles, varying widely among individuals. However, men have been found to showcase more assertive and directive leadership styles. Conversely, women are seen as having more collaborative, democratic, and inclusive leadership styles.

### 3.2.1 Transactional vs. Transformational

Research has indicated that women leaders exhibit more traits of transformational leadership style; they are relationship-oriented, prioritize empowerment, and focus on mentorship. Conversely, male leaders are seen to apply the transactional style of leadership, whereby directives are task-oriented and emanate from performance-based rewards (Krinzman, 2021).

### 3.2.2 Collaborative vs. Directive

Women leaders often stress the importance of a collaborative process of decision-making and inclusive leadership styles. Conversely, men may adopt more directive and authoritative methods. Women are more inclined toward participatory mechanisms that allow diverse perspectives and encourage consensus-building, whereas male leaders are more likely to adhere to hierarchy and top-down decision-making (Krinzman, 2021).

## 3.3 EXPLORING WOMEN'S LEADERSHIP STYLES IN THE DIGITAL WORLD

Within the rapidly evolving landscape of technological companies, leadership styles play a pivotal role in framing organizational culture, promoting innovation at the workplace, and fostering success. As mentioned earlier, the technological industry has always been tagged as male-dominated, the number and influence of female leaders have increased in recent times (Hupfer et al., 2023). Few leadership styles have emerged more prominent over others in women in tech-driven industries. We will explore the most significant ones that have empowered women to foster a culture of inclusivity, success, and growth with their strategic vision, creativity, solution-oriented approaches, and effective communication.

### 3.3.1 Transformational Leadership: Fostering Innovation

Transformational leadership revolves around inspiring and motivating followers to achieve remarkable results by fostering a shared vision, encouraging innovation with creativity, and nurturing individual growth. Women leaders inspire their employees to bring out their best at work. Women have been observed to motivate team members to work beyond their comfort zone (Krinzman, 2021). Their guidance helps employees unleash their full potential and exceed expectations to the point where they bring transformation in the workplace and into their lives.

Transformational leadership style goes beyond transactional management and focuses on helping the organization transition from one phase to another. One such transition occurred in Hewlett Packard Enterprise (HPE) in 2014 under the leadership of its then-CEO, Meg Whitman. From 2013 to 2014, rapid changes in the IT industry were reducing inter-business complementarity. Conversely, customer and competitive forces were driving the intra-business complexities of the consumer-oriented and enterprise-oriented ventures. Witnessing these changes, Whitman decided to split HP into two companies: HP, In. and Hewlett Packard Enterprise (HPE) (Burgelman, 2016). When Whitman took over the role of CEO in 2011, HPE was badly damaged. However, by the time FY 15-16 ended, HP had become two new $50 billion-plus companies, both running successfully (Ignatius, 2016).

Women adopting transformational leadership style bring diverse perspectives, collaborative approaches, and effective communication to the table, creating an environment conducive to innovation. As the CEO of eBay (which was before her role at HPE), Meg Whitman oversaw its transformation from a small online auction site with a handful of employees to a global e-commerce powerhouse. Under her leadership, the company's consumer base surged from a few hundred to millions globally.

Wu et al. (2021) discovered that companies with women as CTOs are more innovative than those with male counterparts. This culture is more prominent for organizations with a stronger innovation-supportive culture or female CEOs. Using mediation analyses, they could validate the transformational leadership style as a logical and possible approach through which women efficiently fostered innovation with creativity in their firms.

Another study showed that organizations with female CEOs are more likely to innovate. It's because of women leaders' intrinsic ability to inspire managers and employees to engage in new, creative, and sustainable business practices (Khushk et al., 2022) transforming the way employees perceive their work, that leads to enhanced motivation and productivity.

#### 3.3.1.1 Exhibiting Empathy through Transformational Leadership

At the heart of transformational leadership lies empathy, the capacity to understand and connect with others on an emotional level. In a tech-driven environment, emotional intelligence becomes a critical leadership skill. Women leaders often excel in this aspect, emphasizing empathy, communication, and relationship-building. Recognizing the importance of understanding and navigating employees' emotions, women leaders create a positive and inclusive work culture that fosters innovation and meaningful change.

Empathy, often characterized as the cornerstone of effective leadership, entails the ability to perceive and understand the emotions, thoughts, and experiences of others. A study conducted by Ayman et al. (2009) found that women leaders are more empathic, humble, and sensitive toward their subordinates than their male counterparts, which plausibly resulted in having a greater tendency for transformational leadership.

Women leaders, who historically have been attributed with higher levels of emotional intelligence, are leveraging empathy as a central tenet of their leadership approach in digital companies. This empathetic leadership style is particularly evident in the transformational leadership model, which emphasizes inspirational motivation, intellectual stimulation, individualized consideration, and idealized influence. Women leaders, in particular, have demonstrated an innate ability to cultivate emotional intelligence and self-awareness, which helps them to find ways to manage stress and anxiety during times of change.

#### 3.3.1.2 Strategies for Women to Implement Transformational Leadership

Women encompassing transformational leadership create a vision regarding the transformation in the organization and openly communicate about it among subordinates. These communications include strategies to transform the company's ethics and work culture. However, foremost, they must convey the necessity to undergo transformation to their employees, for which they need to:

- design a roadmap for reaching the long-run goals of transformation of the organization, while still maintaining a sense of urgency for short-term and quick results
- build influential relationships and coalitions within the company by identifying and immediately addressing resistance or skepticism
- involve the middle management in the transformational processes right from the initial phases, equipping them with due guidance and resources that would enable them to effect changes within their teams
- strategize and pilot change initiatives in smaller areas of the organization before scaling up; and
- identify employees who can advocate for the change initiatives more effectively than others, and communicate with them the tailor-made strategies and goals required to bring about a specific change.

### 3.3.2 Collaborative Leadership: Harnessing Creativity and Connectivity

Women leaders in the digital age often emphasize collaboration and teamwork. In a technology-driven world where cross-functional collaboration is essential, women excel in building inclusive teams that leverage diverse skills and perspectives. The ability to foster collaboration harnesses connectivity, creativity, problem-solving, and adaptability—essential qualities in rapidly changing technological landscapes. This leadership style becomes more critical when a company's management decides to opt for mergers and acquisitions for its growth and cement its place in the market.

Women are equipped to exhibit a collaborative leadership style due to a general awareness of others. They actively listen, comprehend, and respect their subordinates' perspectives, thereby manufacturing an environment of trust and mutual respect and leading to improved teamwork. A Salesforce study showed that 86% of employees and executives identified the lack of collaboration as a primary cause of workplace failures (Chattleworth, 2023). Collaborative leadership leads to enhanced and well-informed decision-making as it benefits from using the team's collective intelligence. A study done by the Institute for Corporate Productivity (i4cp) and Rob Cross, Edward A. Madden Professor of Global Business at Babson College, examined 1100 companies. They found that companies encouraging collaborations showcased a five-times better performance (Gaskell, 2017). Furthermore, these companies were found to cultivate innovation and creativity by promoting diverse perspectives and ideas.

#### 3.3.2.1 Promoting Inclusivity through Collaborative Leadership

Women exhibiting a collaborative leadership style foster a culture of conviction and inclusivity, leading to a more equitable workplace. By recognizing diverse perspectives and encouraging inclusive coalitions, employees feel heard, valued, and included in decision-making processes. This approach, in a study by the Peterson Institute for International Economics, has been found to increase profitability by 15% (Chattlesworth, 2023). Tech companies stand on pillars of innovation and creativity. Women promoting inclusivity through collaborative leadership strive to strengthen these pillars. Women leaders champion inclusive innovation by recognizing the importance of diverse voices in the creative process. In technological firms, where innovation is at its forefront, diverse teams with varied perspectives contribute to more comprehensive problem-solving and product development.

Data by Development Dimensions International, a global management consulting company highlights that organizations with 30% or more women in senior roles are 1.9 times more likely to be rated as having an inclusive culture than those with fewer women (Byham et al., 2024). By actively promoting diversity and inclusion, women leaders create safer workplace environments where everyone feels empowered to contribute their unique insights, leading to increased productivity.

In the digital era, an inclusive collaborative leadership style is proving to be significant in unleashing existing creative talent and driving sustainable growth. Megan Smith, as the former U.S. Chief Technology Officer under the Obama administration, championed initiatives to promote collaboration between government agencies, private companies, and non-profit organizations to tackle complex challenges such as cybersecurity and digital inclusion. Her strategies led to forming partnerships such as the Tech Jobs Tour, which aimed to diversify the tech workforce by providing training and opportunities to underrepresented groups (Arnold, 2017). The Tech Jobs Tour in 2017 saw 140+ employers committed to hiring diverse talent across the USA (Tech Jobs Tour, 2017). Her collaborative efforts resulted in tangible outcomes, benefiting the tech industry and society at large. It is because the tour enabled companies to hire a diverse pool of talent with unconventional mindsets and empowered people from marginalized groups to connect to their nascent tech sectors locally.

#### 3.3.2.2 Strategies for Implementing Collaborative Leadership

Championing collaboration with inclusivity creates a working environment where teams thrive to achieve collective success. To successfully implement collaborative leadership, women leaders foster an environment of trust and mutual respect, for which they:

- Maintain a culture of belonging, wherein all individuals are treated with dignity, making them feel appreciated and empowered to share their unique prospects and talents
- Promote diversity in teams by fostering inclusive hiring practices and providing equal opportunities for advancement
- Foster open communication to maintain transparency and trust
- Remain receptive to diverse perspectives and constructive feedback
- Encourage risk-taking and innovation, empowering teams to take ownership of their work
- Build strong relationships with team members and stakeholders; and
- Conduct team-building activities and recognize individual contributions.

### 3.3.3 Adaptive Leadership: Constantly Evolving

A leader's response to change determines the organization's future. One who can adapt swiftly to changes can surely bring her organization to survive and thrive. Women leaders have demonstrated a remarkable ability to adapt to the fast-paced nature of the digital world. Every month, a new technology comes up on the market, improving a product or service significantly. The constant evolution of technology requires leaders who can quickly pivot, learn new skills, and guide their teams through change. Women, known for their adaptability, excel in navigating uncertainties and steering organizations toward innovation.

Ginni Rometty's tenure as the CEO of IBM exemplifies adaptive leadership in the face of industry transformation. Under her leadership, IBM embraced emerging technologies like cloud computing and artificial intelligence while simultaneously restructuring the company to stay competitive in a rapidly evolving market. Rometty launched a "Strategic Imperatives" initiative, which focused on developing high-growth areas such as cloud, analytics, security, and upgrading products and services under these segments. Consequently, income from these segments contributed 46% of IBM's total revenue in 2017, enabling the company to adapt to changing market demands and regain its position as a leader in the tech industry.

#### 3.3.3.1 Adaptive Leadership Framework for Women in Tech

Women leaders are adept at embracing change rather than resisting it. They encourage experimentation, adaptability, and a growth mindset among their teams. In tech companies, women adopting the adaptive leadership style must reflect on fundamental

## TRANSFORMATIONAL VS. COLLABORATIVE LEADERSHIP STYLE

Certain facets of transformational and collaborative leadership might seem to overlap. However, there are notable differences in the adoption of these leadership styles (See Table 3.1), which also require leaders (regardless of their gender) to act differently and predominantly exhibit different attributes. Both styles are effective in driving organizational success, but they differ in their approaches to leadership and their emphasis on individual versus collective dynamics within teams. In summary, while transformational leadership emphasizes visionary inspiration and individualized consideration, collaborative leadership prioritizes teamwork, facilitation, and shared decision-making.

**TABLE 3.1**
**Differences in the adoption of Transformational and Collaborative Leadership**

| | Transformational Leadership | Collaborative Leadership |
|---|---|---|
| Fostering Innovation | Through articulating a compelling vision and inspiring employees to grow personally as the organization progresses | By creating an environment conducive to cross-functional teams working together and having intellectual discussions with one another |
| Inclusivity | Not necessary | Integral to collaboration leadership style |
| Intrinsic personality requirement of a leader | To be charismatic and visionary | To be democratic |
| Critical attribute or trait required | Empathy and emotional intelligence | Ability to harness connectivity among employees from different sections of the organization |
| The type of management required the most | Change Management | Consensus Building and Team Building |
| Best suited for situations like | Undergoing crisis | Undergoing merger or acquisition |

aspects of change and the impact of change on the employees within the organization. At the same time, following the six principles stated by Heifetz and Laurie (1997), they need to:

- **Get on the Balcony:** Women must have a larger view of the market and movements within the organization as if they are standing on a balcony above the field of the play. This will help them see the context for change or create one. Also, it will help them identify the struggles of their employees over values and power, and other functional and dysfunctional reactions to change
- **Identify the Adaptive Challenge:** Leaders must differentiate a challenge that requires a change of norms, employee roles, and organizational processes from the ones that are technical in nature. Technical problems can be assigned to an expert team, who can then solve them following the existing norms (See Figure 3.1)
- **Regulate Distress:** It requires a leader to expose her employees to existing conflict and let them feel external pressures to an extent they can withstand the stress
- **Maintain Disciplined Attention:** When people in executive teams or at middle managerial levels show work avoidance behavior due to ego (especially seen in the case of male subordinates with women being the leader) or incompetencies, women leaders must get their employees to confront tough trade-offs in values, procedures, operating styles, and power
- **Give the Work Back to People:** Leaders must provide a clear direction for change, after which they need to give them the freedom to take the initiative

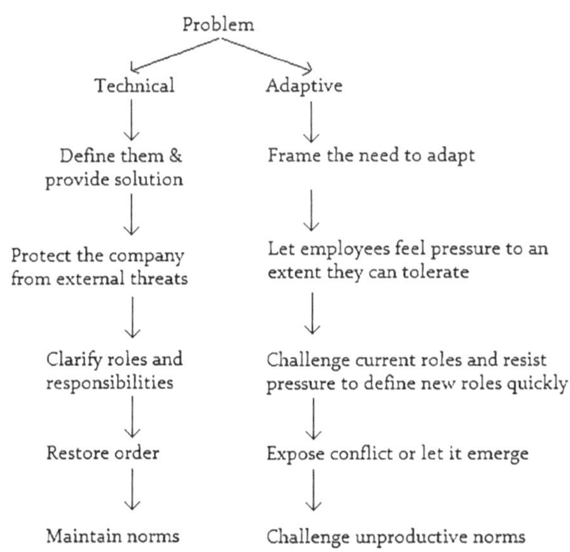


**FIGURE 3.1** Differences in dealing with a technical and adaptive problem using different leadership styles and following different mannerisms. (Adapted from Heifetz and Laurie, 1997.)

in solving problems. Simultaneously, they need to make their managers and employees responsible for the work, which contributes toward the change

- **Protect Voices of Leadership from Below:** Leaders must work with humility and reflect on opinions shared by people at lower levels, especially when they happen to be experts at something specific. It is because they might have a deep valuable insight into the functionality of a product or service, which a leader might not have access to.

### 3.3.4 Ethical Leadership: Building Trust

As technology advances, ethical considerations become increasingly important. Any digital product or service, not limited to those requiring the internet or the use of a chip, has a risk of privacy breach and data leakage. Ethical leadership is crucial for maintaining trust, safeguarding privacy, and promoting equity in the digital world, where decisions can have far-reaching consequences. Women leaders may often prioritize ethical decision-making, demonstrating a commitment to responsible and sustainable practices. Lisa Su, the CEO of Advanced Micro Devices (AMD), has emphasized data security and privacy in developing AMD's processors and AI-powered computers (Carlson, 2024), reflecting a commitment to ethical considerations in technology design. By prioritizing these principles, she helped AMD earn trust and credibility among consumers and stakeholders.

#### 3.3.4.1 Key Principles of Ethical Leadership

Ethical leadership goes beyond mere adherence to rules; it involves a commitment to values, integrity, and social responsibility.

- **Values-Driven Decision-Making:** Ethical leaders prioritize integrity, transparency, and respect in their decision-making processes, ensuring alignment with organizational values and societal expectations
- **Stakeholder Engagement:** They actively engage with stakeholders, including employees, customers, and communities, to understand their concerns and perspectives, fostering trust and collaboration
- **Accountability and Transparency:** Ethical leaders hold themselves and their organizations accountable for their actions, openly communicating about ethical dilemmas and decisions, and taking responsibility for addressing any ethical lapses. It helps them build brand reputation and integrity in an era where corporate responsibility is under constant scrutiny.

### 3.3.5 Laissez-Faire Leadership: Empowering Employees

The laissez-faire leadership style calls for a very minimal level of interference from the leader to maintain a hands-off approach, offering employees much freedom and autonomy to pursue projects and make decisions independently. Generally speaking, laissez-faire leaders empower their teams with the required toolsets or information for assigned projects. Women leaders adopting this style understand quite well how creative ideas take birth, ultimately manifesting into innovative products, which is the utmost requirement of tech-driven industries. They never assume authority over the subordinates and step aside after providing necessary resources while leaving their team members free to plan, organize, make decisions, solve problems, and finish the task given to them.

In the tech industry, where creativity and experimentation are paramount, laissez-faire leadership can catalyze breakthrough innovations and foster a culture of ownership and empowerment. Susan Wojcicki, the former CEO of YouTube, was known to provide her subordinates with enough recreational opportunities, elasticity in work shifts, and creative liberty that helped them sustain a high level of creativity. Under her leadership as a product manager and later as a senior vice president of Advertising & Commerce at Google, the company saw development and major updates of AdSense and AdWords. By the time Wojcicki transitioned to a new role as the CEO of YouTube in 2014, Google was generating more than $50 billion yearly through advertising alone (Bianchi, 2024).

#### 3.3.5.1 Key Strategies of Laissez-faire Leadership

The laissez-faire approach particularly works well within an organization where employees are creative, skilled, and self-motivated. Women leaders in tech companies can empower their teams through laissez-faire leadership by:

- Granting them autonomy to make decisions, pursue projects, and explore innovative ideas independently
- Encouraging them to think outside the box and explore unconventional solutions to complex problems; and
- Creating a culture that embraces risk-taking and failure as essential components of the innovation process, supports and encourages employees to take calculated risks, and learns from setbacks.

### A BRIEF NOTE ON WOMEN LEADERS WITH STEM BACKGROUND

While technical proficiency is not exclusive to any gender, women leaders in the digital age are showcasing their tech-savvy capabilities in the areas of science, technology, engineering, and mathematics (STEM). Many have backgrounds in STEM fields and leverage their expertise to make informed decisions. This technical proficiency not only instills confidence in team members but also positions women leaders as credible advocates for technology-driven strategies. Ursula Burns, who served as the CEO of Xerox from 2009 to 2016, did her Bachelor's and Master's degrees in Mechanical Engineering. She became the first African-American woman to lead a Fortune 500 company. Her background in STEM played a significant role in her leadership, guiding Xerox through a transformation into a services-led technology company by acquiring Affiliated Computer Services (ACS), an outsourcing company (Bryant, 2010). A total of 96% of Xerox shareholders along with 86% of ACS shareholders voted in favor of the acquisition. Consequently, Xerox's total market opportunity rose to $500 billion, and its employee base grew to 130,000 (Burns, 2021).

Anne Wojcicki is the co-founder and CEO of 23andMe, a personal genomics and biotechnology company. She earned a Bachelor of Science degree in Biology from Yale University. Her STEM background empowered her to steer her company toward creating the first personal genetic test with FDA authorization to send genetic variant information directly to consumers (Sauer, 2023).

Coming from a different background, Marry Barra, the Chairperson and CEO of General Motors (GM) holds a Bachelor of Science degree in Electrical Engineering. She is the first woman to lead a major global automaker. Her STEM background and leadership qualities have been instrumental in guiding GM through advancements in electric and autonomous vehicles, as well as digital innovation in the automotive industry, putting the company on the map as a major player in leading the transition toward greener and more sustainable forms of transportation (Swallow, 2023).

## 3.4 CASE EXAMPLES

It is apparent that the digital landscape incessantly changes at a rapid pace. Women leading organizations must be on their toes to adapt to the change. But they cannot do it if they stick to one style of leadership showcasing a specific set of attributes. A woman can successfully lead if she inculcates characteristics from different leadership styles at different times. Successful leaders have one prominent style that prevails over the others. In the case of women, it is transformational, collaborative, laissez-faire, or adaptive. However, they keep adding different attributes to their armory and use them when needed. For instance, when the organization is in crisis or its growth is stagnant, women leaders must show more transformational leadership traits. Similarly, if the organization is going to acquire small companies for growth or if the firm has earned a project that involves multi-domain expertise, its leader must adapt to exhibiting collaborative leadership. Hence, it is critical to explore real case examples of women who helped their tech companies grow to prominence by practically showcasing a blend of different leadership styles.

### 3.4.1 Growth and Transformation of SHEROES under Sairee Chahal's Leadership

Merger or acquisition (M&A) creates an environment of uncertainty, complexity, and ambiguity for all those involved. A multitude of risks accompanies mergers and acquisitions. Potentially one of the greatest fears shared by employees is the fear of layoffs, which can lead to negatively impacting employees' productivity. However, SHEROES, under the leadership of Sairee Chahal, completed multiple M&As successfully and efficiently. In a period spanning from 2016–2020, SHEROES acquired six niche women-centric companies and grew to 20+ million women users by 2020 (Kumari and Dash, 2021). SHEROES was founded by Chahal in 2013 as a community for women seeking jobs and career opportunities. With SHEROES, Sairee

Chahal designed innovative pathways to build a favorable and supportive ecosystem for women to work in India, counting massively on technology to solve the issues of the gender gap in the workforce.

Within a few years of its launch, Chahal, an ASPEN Leadership Fellow, began forming partnerships with other start-ups offering varying services for women's welfare. Consequently, it helped SHEROES transform into launching new communities that discussed arts, health, and relationships, allowing women to share their life stories and achievements. In 2016, the company acquired Gharkamai.com, an online platform that helped stay-at-home professional women find work, and LoveDoctor.in, a relationship and sexual health counseling platform (Sharma, 2016). Chahal, as a collaborative leader, ensured participation from representatives from both its acquired organizations. Cross-organizational teams were formed with a common vision, fostering a culture of creativity and problem-solving within SHEROES.

Further, in August 2017, SHEROES acquired Women Restart, a platform aiming to help women restart their careers after a break (Paul, 2017). These acquisitions enabled the company to strengthen its communities of women in their personal and professional pursuits by enhancing their offerings with career counseling, empathetic relationship counseling, and training services (Daswani, 2017).

By 2017, SHEROES was offering a plethora of services for women to engage in, including freelancing and internship programs and providing job opportunities (both part-time and full-time) to work in corporations. Consequently, by the beginning of FY 17-18, SHEROES had become India's largest community of women, supported by an ecosystem of 300 partners and over 17,000 companies. Of course, every time a company was acquired, its senior leaders and founders were retained and given top executive positions in SHEROES. Lay-offs were avoided, sending a message of security amongst employees. For instance, Sheetal Arora, founder of Women Restart, was appointed as Director of Research and Resources in SHEROES (Paul, 2017).

Carrying the vision forward of touching the lives of 100 million women by the end of FY 24-25, Chahal continued her streak of acquiring startups. Subsequently, she acquired Babygogo in December 2017, Maya in January 2019, and Naaree.com in May 2020 amidst the lockdown. It allowed SHEROES to strengthen its offerings in the field of baby care, parental services, self-care, and health for its community.

#### 3.4.1.1 Forming Strategic Alliances

SHEROES, on July 21, 2020, formed a strategic alliance with Otipy, a community-based platform to connect consumers, resellers, and farmers to bring fresh produce to their doorstep. The partnership between Otipy and SHEROES was planned to be based on a revenue-sharing model, with strong synergies between both app-based platforms (BW Disrupt, 2020). Chahal strategically forged partnerships with organizations that aligned with SHEROES' mission. Collaborations with companies, NGOs, and educational institutions expanded the platform's reach and enhanced the range of opportunities available to SHEROES members.

#### 3.4.1.2 User-Centric Design

The success of SHEROES can also be credited to its user-centric design. Chahal actively listened to the needs of the community and adapted the platform to cater to evolving requirements. This iterative approach ensured that SHEROES remained a relevant and impactful resource for women seeking support and empowerment. Chahal built SHEROES in a way that empowered its women users to leverage technology to build great products, grow their ventures, and market themselves better, without too many overheads. Women could leverage the platform to market their services and products, build their brand story and reputation, launch crowdfunding campaigns, and much more. The platform allowed them to proactively seek support, whether for career counseling or mental health guidance.

Sairee Chahal's leadership at SHEROES exemplified a visionary approach to empowering women through community, technology, and strategic partnerships. By fostering inclusivity, embracing technology, and promoting an entrepreneurial mindset, Chahal created a platform that not only addressed the unique challenges faced by women but also propelled them toward personal and professional success. SHEROES, under Sairee Chahal's guidance, stands today as a beacon of empowerment, illustrating the transformative impact of visionary and collaborative leadership in the realm of women's advocacy and support.

### 3.4.2 Steering Gravity Sketch to a $59 Million Company—Exploring the Leadership Style of Daniela Paredes

Daniela Paredes is one of the co-founders of Gravity Sketch, a fast-growing UK technology start-up founded in 2014. The co-founders developed a next-generation 3D design tool as a sketching application. Contrary to existing CAD-based models and traditional pen-and-paper drawings, Gravity Sketch designers used a headset and controllers to engage in their creative process and the space they were working on (Women in Tech, 2019). Paredes oversaw the company's daily operations and developed growth strategies for future markets like education. She worked closely with designers using Gravity Sketch daily and understood how the tool integrated seamlessly into a user's workflow. Having graduated in Innovation Design Engineering from The Royal College of Art and Imperial College London, she kept giving her expert opinion on how the product could be improved and developed further. Consequently, she developed strategies and processes that helped the company build tools and mechanisms that enabled users to get the best out of the software and meet their objectives (Women in Tech, 2019).

In 2019, Paredes was one of the nine women recognized in Innovate UK's Women in Innovation Awards and won £50,000 in funding as a grant (Evanson, 2019). Adapting to the unprecedented shift in the work culture globally, her visionary and collaborative leadership paved the path for Gravity Sketch to become a cross-platform during the COVID-19 pandemic. It allowed designers and their teams to be in the same virtual space along with customers at every step of the product development. Stakeholders from different disciplines could collaborate, develop designs, share feedback, and discuss issues at every stage of product development (Nailing the bottom-up sales, 2022).

In September 2020, Paredes helped the company raise $3.73 million as seed capital (Marzook, 2020). By that time, companies like Ford, Nissan, and Reebok, along with 50,000+ creative professionals were using the immersive platform (Marzook, 2020). Following the bottom-up sales approach, the company garnered one million downloads by FY 21-22. With a community of 100,000+ monthly users, the company recorded $6 million in revenue in FY 21-22, which was 4 times the previous year (Paul, 2022). Moreover, in April 2022, the UK-based startup raised $33 million in the Series A round of funding. The investment enabled the company to accelerate product development and transform long-neglected digital design workflows, empowering it to further tap into the world of the metaverse (Paul, 2022).

### 3.4.3 Becoming Amazon Basics of Cattle Trading—Decoding the Journey of Neetu Yadav and Kirti Jangra

Co-founded by Neetu Yadav, Kirti Jangra, and two other colleagues, "Animall" was launched in November 2019 to facilitate dairy farmers to trade cattle online while offering healthcare and well-being services for cattle heads. The company launched the app as a peer-to-peer (P2P) platform for cattle trading (Singh, 2021; Jain, 2022), which aimed to serve as a farmer-centric tool offering a technology-enabled ecosystem to cattle farmers, wherein they could access the market with a tap of their mobile phones. The vision of Yadav and Jangra was to create a digital version of the animal fair—the route farmers usually took for cattle trading. With this P2P app, Animall could provide dairy farmers with the similar ease and variety of choices that consumers of mainstream goods experience while shopping for daily stuff. They now had access to more choices of cattle, transparency, and availability of information, but without the involvement of brokers in striking a deal.

In March 2021, Animall scored a fresh round of funding, garnering approximately $2 million at a valuation of $27 million, and claimed to be the largest trading platform for cattle (Vardhan, 2021). In July 2021, a major Series B round of approx. USD 13.8 million took place at a valuation of $70.6 million (Upadhyaya and Tyagi, 2021). By the end of FY 20-21, more than four million people had downloaded the app in three states of India (Yadav, 2021). In a little under two years, the groundwork done by Jangra and Yadav saw Animall selling over 500,000 cattle, amounting to $320 million of GTV (gross transaction value) (Singh, 2021). By December 2021, the duo had overseen the registration of 180,000 cattle for sale on the app, establishing over one million buyer-seller contacts (Jain, 2022). By the time FY 22 ended, the company had reached out to more than eight million dairy farmers and facilitated the sales of 850,000 cattle (Ahaskar, 2022).

Although Animall started with the P2P model, the leadership of Jangra and Yadav pivoted the company to work on the marketplace model with inventory in late 2021 by offering Animall-labelled and Animall-verified cows and buffaloes (Saha, 2023). The idea was to add more value to the cattle buyer by eliminating misinformation from the seller, which, at times, would leave the buyer with cattle of low quality. To retain the trust of its users, Jangra and Yadav decided to offer cattle verified by the company and bearing a stamp of quality. This possibility of buying verified cattle directly from the company created a favorable transaction space for the seller, so much so that the company's revenue jumped over 4X to INR 30.4 crore in FY23 from 7.18 crore in FY22 (Ashrafi, 2023).

Yadav and Jangra had to adapt their operations and deal with new challenges of logistics and inventory to execute their plan to become "Amazon Basics" for cattle trading, wherein they could create a network of certified sellers from whom any customer could buy through the app (Masters' Union, 2022). The attributes of different leadership styles they showed helped them focus solely on expanding vertically, resolve any conflict or indecision in the boardroom, and streamline operations on the ground. The sheer passion of Yadav and Jangra towards the dairy farming industry inspired them to steer Animall towards becoming an institutional player, wherein the company could standardize cattle prices through SOPs and technology, depending on the stage an animal is in its lifecycle and other factors.

## 3.5 DISCUSSION

The leadership style plays a crucial role in developing effectiveness and determining outcomes of a leader in different settings. There is no one-size-fits-all approach to leadership, and different scenarios call for distinct leadership styles. When leaders understand the sturdiness, unique features, and potential drawbacks of different leadership styles, they are able to adapt their mannerisms and maximize their influence on their teams and organizations. Women, by learning and assimilating a range of attributes of different leadership styles, have carried out certain practices of leadership that have been extremely effective in bringing tremendous success as well as significantly improving the working environment. As Chief Executive Officers, Chief Technology Officers, or simply as heads of departments, their style of leadership has helped steer companies through the digital age while encouraging creativity and pushing boundaries. Ginni Rometty at IBM and Marry Barra at General Motors have proved the arms of women in technology improvement, flaunting women's pivotal roles in driving tech giants toward future success.

Women as business leaders are driving innovation through startups and tech ventures. They are doing so by encouraging the development of disruptive technologies and challenging industry standards. Owing to the increasing boulders of female entrepreneurs starting their ventures in the fields of artificial intelligence (AI), biotech, and fintech, women are changing the face of the technology industry and finding their own places in the digital economy. Women leaders, no matter what leadership approach they use, have to master a collection of skills and inherent traits that can motivate change in an organization and inspire employees to perform in alignment with their vision. To achieve that, they must stay true to their humanity and values. It will then allow them to identify and embrace the opportunities for personal and organizational growth, and consequently, navigate the challenges of driving meaningful and lasting change within their organizations.

## 3.6 CONCLUSION

In conclusion, women leaders in the era of technology and digitalization bring a blend of collaborative, adaptable, and transformational leadership styles. Their emphasis on ethical decision-making and a commitment to continuous learning sets the stage for success in a rapidly evolving digital landscape. Women have a natural knack for showcasing empathy and excelling in communication. These qualities enable them to guide their employees by promoting mentorship; transfer authority to teams by emphasizing empowerment; and build relationships among managers and subordinates. In a tech-driven environment, where the workplace environment often gets robotic, the emotional intelligence of women leaders comes in handy to create a progressive work culture that fosters meaningful change in their organizations. They are competent in translating their empathy into inspiring their managers and other employees to bring transformation and embrace change. As a fundamental component of their leadership, empathy helps them relate to, comprehend, and connect with diverse teams, which promotes inclusivity and collaboration. Their adaptive approach enables them to navigate the rapidly evolving digital landscape, seizing opportunities and mitigating risks. From driving innovation and leading strategic initiatives to advocating for diversity and inclusion, women are driving positive change in the digital landscape. As women continue to break barriers, their unique approaches continue to not only enrich the tech industry but also pave the way for a more equitable and socially responsible digital future.

## KEY TERMS AND DEFINITIONS

- **Leadership style:** Leadership style encapsulates the distinct approach a leader adopts in directing, managing, and inspiring their team or organization towards a common goal. Leadership style pertains to the manner in which a leader interacts with their followers and executes their role. A leadership style is characterized by different methods of decision-making, communication, and motivation.
- **Leadership traits:** Leadership traits refer to the inherent qualities, characteristics, and attributes that distinguish effective leaders from others. While leadership styles are influenced by situational factors and can be developed over time, leadership traits are often regarded as the inherent characteristics and qualities that enable individuals to lead effectively. These qualities are innate to an individual's personality.
- **Inclusivity at Workplace:** Inclusivity in the workplace refers to the deliberate and proactive effort to create an environment where every individual, regardless of their background, identity, or characteristics, feels valued, respected, and empowered to contribute their unique perspectives and talents. It involves embracing diversity in all its forms and fostering a culture of equity, belonging, and mutual respect among employees.
- **Work ethics:** Work ethics refers to the set of moral principles, values, and behaviors that guide individuals in their professional conduct and interactions within the workplace. It encompasses attitudes, habits, and standards of behavior that govern how individuals approach their work, fulfill their responsibilities, and interact with colleagues, clients, and stakeholders. Work ethics reflect a commitment to integrity, professionalism, accountability, and excellence in all aspects of work-related activities.

- **Workplace culture:** Also called organizational culture, workplace culture is defined by the set of cumulative interactions between people (at any level or position) within a facility. These interactions are driven by individuals' shared values, beliefs, norms, and practices, reflecting on organizational workflow, cultural norms, and how decisions are made, and relationships are formed among employees.
- **Gross Transaction Value:** A financial indicator, Gross Transaction Value (GTV) shows the total amount of all transactions performed during a given time, usually within a business, platform, or financial system. It includes the total amount of all purchases, sales, and other financial transactions, less any amounts deducted for discounts, taxes, fees, or refunds.
- **Bottom-up sales approach:** The bottom-up sales approach is a strategy employed by businesses to grow their sales by focusing on individual customers or small accounts, and gradually building up to larger sales opportunities. In contrast to the traditional top-down approach, which typically targets larger accounts or customers first, the bottom-up approach begins with smaller, more manageable sales transactions and gradually expands the customer base over time.
- **P2P platform:** A peer-to-peer (P2P) platform is a type of online marketplace or network that facilitates direct transactions between individuals or entities, without the need for intermediaries such as traditional institutions or centralized authorities. In a P2P platform, users interact directly with one another to buy, sell, rent, or exchange goods, services, or assets, leveraging technology to facilitate transactions, communication, and trust-building among participants.

## REFERENCES

Ahaskar, A. (2022). "Cattle trade and livestock farming in India get a touch of AI, IoT." *Techcircle*. www.techcircle.in/2022/04/06/cattle-trade-and-livestock-farming-in-india-get-a-touch-of-ai-iot

Arnold, J. (2017). "First female CTO of the U.S. Megan Smith hopes Tech's history can repeat itself." *Forbes*. www.forbes.com/sites/jennaarnold/2017/11/08/techs-history-hopefully-can-repeat-itself/?sh=1d66a28e4b5a

Ashrafi, M.S. (2023). "Animall's revenue shoots up 4X to Rs 30 Cr in FY23." *ENTRACKR*. https://entrackr.com/2023/11/animalls-revenue-shoots-up-4x-to-rs-30-cr-in-fy23

Ayman, R., Korabik, K., & Morris, S. (2009). "Is transformational leadership always perceived as effective? Male subordinates' devaluation of female transformational leaders." *Journal of Applied Social Psychology, 39*(4), 852–879. https://doi.org/10.1111/j.1559-1816.2009.00463.x

Bianchi, T. (2024). "Advertising revenue of Google from 2001 to 2023." *Statista*. www.statista.com/statistics/266249/advertising-revenue-of-google/

Bryant, A. (2010). "Ursula Burns, aiming to redefine the xerox culture." *The New York Times*. www.nytimes.com/2010/02/21/business/21xerox.html

Burgelman, R.A. (2016) "Meg Whitman resolves strategic integration challenges, from better together to splitting HP in two." *Becoming Hewlett Packard: Why Strategic Leadership Matters* (pp. 293–328). Oxford University Press. https://doi.org/10.1093/acprof:oso/9780190640446.003.0009,

Burns, U. (2021). "How I saved Xerox from a near-death experience." *Built In*. https://builtin.com/corporate-innovation/xerox-ursula-burns

Bwalya, A. (2023). "Leadership styles." *Journal of Entrepreneurship, Management and Innovation, 11*(8). https://doi.org/10.6084/m9.figshare.23932113

BW Disrupt. Otipy to acquire women resellers through its partnership with SHEROES. (2020, July 21). https://bwdisrupt.com/article/otipy-to-acquire-women-resellers-through-its-partnership-with-sheroes-299618.

Byham, T.M., Neal, S., & Rhyne, R. (2024). "Women in leadership statistics: Insights for inclusion." *DDI.* www.ddiworld.com/blog/women-leadership-statistics

Carlson, K. (2024). "At SXSW, AMD CEO Lisa Su says 'we must go faster' on AI, while keeping a 'watchful eye'." *Austin American-Statesman.* www.statesman.com/story/business/technology/2024/03/12/sxsw-tech-lineup-2024-amd-ceo-lisa-su-future-of-ai-artificial-intelligence/72886020007/

Chattlesworth, A. (2023). "Collaborative leadership a new approach for female leaders." *Medium.* https://medium.com/@atticuschattlesworth/collaborative-leadership-a-new-approach-for-female-leaders-66017cdd3ed9#

Daswani, S. (2017). "Sheroes acquires mother's community startup Babygogo." *INC42.* https://inc42.com/buzz/sheroes-acquires-babygogo/

Dwiri, B., & Okatan, K. (2021). "The impact of gender on leadership styles and leadership effectiveness." *International Journal of Science and Research (IJSR), 10*(1), 1419–1434. 10.21275/SR21126183926.

Evanson, D. (2019). "Inventor of 3D sketching tool named one of the UK's most innovative women." *Imperial.* www.imperial.ac.uk/news/190512/inventor-3d-sketching-tool-named-uks/

Gaskell, A. (2017). "New study finds that collaboration drives workplace performance." *Forbes.* www.forbes.com/sites/adigaskell/2017/06/22/new-study-finds-that-collaboration-drives-workplace-performance/?sh=2c5c74e83d02

"Gravity – Daniela Paredes, Co-founder." *Women in Tech.* 2019. www.womenintech.co.uk/gravity-daniela-paredes-co-founder/

Heifetz, R.A., & Laurie, D.L. (1997). "The work of leadership." *Harvard Business Review, 75*, 124–134.

Hupfer, S., Mazumder, S., & Crossan, G. (2022). "Women in tech are cracking the industry's glass ceiling, achieving double-digit gains in leadership roles." *Deloitte Insights.* www.deloitte.com/global/en/our-thinking/insights/industry/technology/women-tech-leadership.html

Ignatius, A. (2016). "We need to intensify our sense of urgency: An interview with Meg Whitman." *Harvard Business Review.* https://hbr.org/2016/05/we-need-to-intensify-our-sense-of-urgency

Jain, S. (2022). "Neetu Yadav and Kirti Jangra: Bringing the digital revolution to $250 bn cattle market industry." *Forbes India.* www.forbesindia.com/article/30-under-30-2022/neetu-yadav-and-kirti-jangra-bringing-the-digital-revolution-to-250-bn-cattle-market-industry/73473/1

Khushk, A., Zengtian, Z., & Hui, Y. (2022). "Role of female leadership in corporate innovation: A systematic literature review." *Gender in Management: An International Journal, 38*(3), 287–304. https://doi.org/10.1108/gm-01-2022-0028

Krinzman, B. (2021). "A review of workplace leadership styles: Men vs. women." *Helios HR.* www.helioshr.com/blog/2015/06/a-review-of-gender-leadership-styles-common-traits-in-men-vs-women/

Kumari, D., & Dash, A. (2021). "SHEROES: Entrepreneurship by acquisitions and organic growth." *Emerald Emerging Markets Case Studies, 11*, 1–21. 10.1108/EEMCS-09-2020-0350.

Marzook, S. (2020). "Gravity Sketch secures £2.9 million Seed investment led by Kindred Capital." *UK Tech Investment News.* www.uktechnews.info/2020/09/16/gravity-sketch-secure-2-9-million-seed-investment-led-by-kindred-capital/

Masters' Union. (2022, September 7). Fireside Chat | Kirti Jangra, Founder, Animall [Video]. YouTube. www.youtube.com/watch?v=9879jNYM6kU

"Nailing the bottom-up sales approach with Daniela Paredes, co-founder of Gravity Sketch." (2022). *Forward Partners.* www.forwardpartners.com/latest/nailing-the-bottom-up-sales-approach-with-daniela-paredes-co-founder-of-gravity-sketch

Paul, B. (2017). "Sheroes acquires women-focused career platform Women Restart." *VCCIRCLE.* www.vccircle.com/sheroes-acquires-women-focussed-career-platform-women-restart/

Paul, M. (2022). "Disrupting 3D design process, Gravity Sketch swipes off $33 million." *TechEU.* https://tech.eu/2022/04/20/disrupting-3d-design-process-gravity-sketch-swipes-off-33-million/

Saha, P.K. (2023). "Animall, a 'Tinder for cattle,' needs soul-searching." *The Morning Context.* https://themorningcontext.com/internet/animall-a-tinder-for-cattle-needs-soul-searching

Sauer, M. (2023, March 17). "How these 3 sisters grew up to become highly successful CEOs and a doctor: It's helpful to realize the world is often wrong." *CNBC.* www.cnbc.com/2023/03/17/how-wojcicki-sisters-grew-up-to-become-highly-successful-ceos-doctor.html

Sharma, D. (2016). "Sheroes takes over women's job portal Gharkamai." *VCCircle.* www.vccircle.com/sheroes-takes-over-womens-job-portal-gharkamai/

Singh, R. (2021). "How two IITians shunned the herd mentality to sell cattle online." *Forbes India.* www.forbesindia.com/article/take-one-big-story-of-the-day/how-two-iitians-shunned-the-herd-mentality-to-sell-cattle-online/70549/1

Swallow, T. (2023). "Lifetime of achievement: Mary Barra." *EV Magazine.* https://evmagazine.com/articles/lifetime-of-achievement-Mary-Barra

"Tech Jobs Tour 2017: That's a wrap." (2017). *Medium.* https://medium.com/@squad_74501/tech-jobs-tour-2017-highlights-aba090755d39

Upadhyay, H., & Tyagi, G. (2021). "Exclusive: Animall raises over Rs 100 Cr at a valuation of Rs 565 Cr." *ENTRACKR.* https://entrackr.com/2021/07/exclusive-animall-raises-over-rs-100-cr-at-valuation-of-rs-565-cr/

Vardhan, J. (2021). "Exclusive: Animall raises funds led by Rocketship VC; valuation crosses Rs 200 Cr." *ENTRACKR.* https://entrackr.com/2021/03/exclusive-animall-raises-funds-led-by-rocketship-vc-valuation-crosses-rs-200-cr/

Wu, Q., Dbouk, W., Hasan, I., Kobeissi, N., & Kobeissi, N. (2021). "Does gender affect innovation? Evidence from female chief technology officers." *Research Policy, 50*(9). https://doi.org/10.1016/j.respol.2021.104327

Yadav, J. (2021). "Apps now sell cows, but getting a farmer to buy one 'before milking it once' is a challenge." *ThePrint.* https://theprint.in/india/apps-now-sell-cows-but-getting-a-farmer-to-buy-one-before-milking-it-once-is-a-challenge/615324/

# 4 Technopreneurship in the Digital Age

## *Merging Technology and Entrepreneurship for Global Innovation*

*Sushant, Savita Sindhu, Meghna Luthra, and Shruti Vashist*

## 4.1 INTRODUCTION TO TECHNOPRENEURSHIP

Technopreneurship is a convergence of echnology and entrepreneurship, culture, and spirit of the venture; it is considered a new and innovative way of conducting business where technology is modified to produce and deliver innovative products, services, or solutions (Choi, Iftitah and Angela, 2024). Technopreneurship has thus been identified as an important player in the new world economy characterized by rapid technological advancement and globalization. It involves leveraging the newest approaches to discovering new markets, addressing meaningful challenges, and creating viable businesses. In contrast to what may be referred to as conventional entrepreneurship, which may involve operating within pre-existing business models, technopreneurship seeks to leverage technology to transform established markets or even create new ones (Olusegun, Akpoviroro and Adebowale, 2019). At its basal level, technopreneurship is associated with innovation and taking risks. Technopreneurs are not mere businesspeople; they are pioneers with a technical background who can innovate and transform the business world (Milton-Smith, 2003). These innovative leaders understand future trends and the possibility of breaking through technologies like AI/ML, blockchain, Internet of Things (IoT), biotechnology, robotics, etc. These technologies are used to create unique approaches to solving specific problems and the solutions are implemented mostly in healthcare, education, energy, agriculture, and finance, and other fields (Tyagi, Aswathy and Abraham, 2020). The entrepreneurial technology influence does not only reflect in new product and service offerings but also in the emergence of new markets and solutions, along with new jobs.

Technopreneurship is characterized by a significant focus on research and development, among its most important aspects. Technopreneurs heavily commit resources to probing and fine-tuning such technologies to suit different markets (Dolatabadi and Meigounpoory, 2013). It enables them to commit a significant number of resources

DOI: 10.1201/9781003537335-4

to research and development; they devote a lot of time to studying the technological trends of the industry and constantly improving themselves to become more competitive. For example, companies in the fintech niche have been developing new solutions using AI and blockchain technologies to increase protection and make financial services more convenient. In the same way, technopreneurs in the healthcare sector are applying new inventions in biotechnology and analytics to transform conventional diagnostic and therapeutic approaches. Technopreneurship also helps solve global issues. As climate change, scarcity of resources, and epidemics continue to worsen, technopreneurs are using technology to find means to reduce these impacts (Shekhar *et al.*, 2023). For instance, renewable energy startups are using improvements in solar, wind, and battery systems to develop better and cheaper energy solutions. Through IoT and data analytics in agriculture, technopreneurs are enhancing crop productivity and minimizing food wastage, which is crucial in addressing the issue of food insecurity. However, with the emergence of digital platforms and the internet, technopreneurship has become even easier compared to the past (Ab Rahman *et al.*, 2024). The use of information technology products in cloud computing, e-commerce platforms, and marketing tools makes it easier for technopreneurs to expand their business worldwide with little capital investment. This has helped to democratize technology and reduce the hurdles that would have otherwise locked out potential technopreneurs keen on participating in the technology-driven entrepreneurship market (Xiong, 2016). Therefore, Technopreneurship is the coming together of technology and entrepreneurship, whereby innovation and risk-taking lead to the development of the next big thing. It represents a significant driver of economic growth, employment, and innovation while carrying considerable responsibility for the future of industries and society. Technopreneurship creates profitable ventures for technopreneurs and plays a pivotal role in addressing some global problems. Technopreneurship will continually enjoy ever-expanding opportunities as technology remains a key success factor in today's world economy.

## 4.2 THE EVOLUTION OF TECHNOPRENEURSHIP

Technopreneurship development shows a process of diversification of technologies and entrepreneurial activity based on improving innovations, social demands, and opportunities in the market. It can be divided into several phases characterized by certain key technological advances and changes in organizations' operations.

### 4.2.1 Pre-Industrial Era (Before 18th Century)

- **Technological Foundations:** Historically, entrepreneurship was mainly focused on hand-made work, commerce, and, in some cases, a small-scale manufacturing process. Technological advances were incremental and were largely concentrated in the areas of agriculture, textiles, and transport
- **Innovations in Tools and Machinery:** Basic technologies such as the plow and wheel were invented, and then advanced technologies such as the printing press were founded on these basic innovations. In this period, inventors focused on making traditional labor more efficient (Diantoro *et al.*, 2023).

### 4.2.2 First Industrial Revolution (Late 18th to Early 19th Century)

- **Mechanization and Industrialization:** The First Industrial Revolution involved a transition of labor from manual labor to machines. Prominent innovations included the steam engine, mechanized textile machinery, and metallurgy. Progressions affected change in the manufacturing, mining, and transport industries
- **Rise of Entrepreneurial Ventures:** Businesspeople started adopting these technologies to establish enormous enterprises. James Watt introduced the steam engine, while Richard Arkwright created the spinning frame, both pioneering technological advancements that effectively set the advancement of industrial capitalism
- **The Emergence of Factories:** The invention of factories, such as the centralization of production units, created a boost in production rates. People tried to make the best use of these technologies, creating businesses and industries and consequently expanding cities (Adeoti, 2019).

### 4.2.3 Second Industrial Revolution (Late 19th to Early 20th Century)

- **Technological Breakthroughs:** The Second Industrial Revolution was spearheaded by emergent technologies like electricity, the internal combustion engine, and improvements in the chemical, steel, and electrical sectors. The advancement of this period was also remarkable as it saw the development of crucial tools such as the telephone, the automobile, and the airplane
- **Rise of Mass Production:** Prominent capitalist figures such as Henry Ford established an assembly line in production, which reduced manufacturers' costs to reasonable prices. This era was characterized by the expansion of powerful industrial business entities that used technology within the context of business strategies to muscle the world markets
- **Technopreneurship Expands:** Some of these are Thomas Edison, who invented the electric lightbulb and used his invention to create a large company, and Nikola Tesla, who developed AC electricity and used it to establish a similar companythis period featured capitalists who incorporated technology with management strategies that embraced product mass production and global marketing (Adeoti, 2019; Ndagi and Salihu, 2019).

### 4.2.4 Post-World War II and the Third Industrial Revolution (Mid-20th Century to Late 20th Century)

- **Technological Innovation in Electronics:** The third Industrial Revolution came in the middle of the 20th century and is widely known as the Digital Revolution. Adopting advanced electronics, telecommunications, and computing technologies propelled it. Technologically, the discovery and invention of the transistor and the evolution of computers and telecommunications systems define modern technopreneurship

- **Birth of the Technology Industry:** Bill Gates (Microsoft), Steve Jobs (Apple), and Larry Page and Sergey Brin (Google) were innovators who formed the tech titans of the present day through personal computing, operating systems, and search engines that revolutionized work, communication, and information access
- **Rise of Silicon Valley:** Technopreneurship hub that further developed in Silicon Valley as startups and technological companies were specifically established to own a focus on semiconductors, computer hardware, and software. Venture capital as a source of funds and the culture of innovation led to the technology industry's growth
- **Internet Boom:** Technological advancement in the 1990s also facilitated the commercialization of the internet, which changed technopreneurship. Julie Wrigley (Tongue-tied) and Bob Parsons (GoDaddy.com) took advantage of the internet to produce novel ideas that form the basis of e-business and web-based organizations (Woo-Je, 2024).

### 4.2.5 Dot-Com Era and Early 21st Century (1990s to Early 2000s)

- **Dot-Com Boom and Bust:** Another defining period can be termed the Dot-Com Boom, characterized by a March of internet-based companies in the late 1990s. Businesspeople rushed into the internet domain, establishing organizations aimed at internet selling, advertising, and communication. Yet, most of these organizations were merely a mirage with unhealthy business models; this led to the infamous Dot-Com Bust in 2000
- **Survivors and Market Shifts:** Nonetheless, several firms such as Amazon, eBay, and Google emerged from the bust and expanded, defining the e-commerce and digital marketing era that followed. These companies grew into corporations with billions of dollars in revenues and motivated a new generation of technopreneurs
- It was in the early 2000s when the phenomenon known as Web 2.0 focused on accessibility, personal contributions, social networking, and collaboration. Entrepreneurs like Mark Zuckerberg (Facebook), Jack Dorsey (Twitter), and Reid Hoffman (LinkedIn) leveraged these trends to build highly successful social networking platforms (Milton-Smith, 2003).

### 4.2.6 The Rise of Mobile and Cloud Technologies (2010s)

- **Mobile Revolution:** The liberalization of smartphones and mobile application services in 21st century opened a new frontier for technopreneurs. Businesspeople leveraged the mobile market, resulting in the development of applications such as Instagram, WhatsApp, and Uber, which altered communication, social networks, and transportation systems
- **Cloud Computing and SaaS:** This was driven by the emergence of cloud computing, which enabled organizations to provide software as a service (SaaS). Some apparent changes were the rise of SAAS companies such as Salesforce.com or Dropbox, bringing technology closer to businesses

- **Sharing Economy:** They also adapted to the sharing economy, which aims to access and share assets and services. Firms like Airbnb and Uber altered conventional industries, such as the hotel and transportation industries, and showed how technology would redefine existing markets (Dada, 2021).

### 4.2.7 Fourth Industrial Revolution (Late 2010s to Present)

- **Artificial Intelligence (AI) and Machine Learning:** The Fourth Industrial Revolution is defined by incorporating advanced technologies such as artificial intelligence, machine learning, robotics, and data science into different sectors. AI is increasingly a feature of startups as young companies create solutions across the healthcare and automobile industries. Fintech, healthtech, and agritech are some sectors where technopreneurs use AI systems to develop novel concepts and solutions
- **Internet of Things (IoT):** Technopreneurs have seized new opportunities arising from integrating IoT devices linking the physical and the digital world in areas like smart homes, connected cars, and smart industries. These technologies are revolutionizing industries such as manufacturing, health, and energy
- **Blockchain and Cryptocurrencies:** Blockchain technology has transformed industries by allowing for distributed systems, is secure, and provides transparency in financial and monetary transactions. The new FinTech companies focus on cryptocurrencies, decentralized finance, and smart contracts to transform the financial industry
- **Augmented Reality (AR) and Virtual Reality (VR):** AR and VR are being used by technopreneurs to deliver engaging content in various fields, such as games, education, health, and the property market. These technologies enable users to interact with digital systems more sociable and creatively (Adeoti, 2019).

### 4.2.8 Technopreneurship in the Post-Pandemic Era (2020s)

- **Acceleration of Digital Transformation:** The COVID-19 pandemic further bolstered the popularization of digital solutions across industries, be it communication platforms (e.g. Zoom, Slack) or e-commerce platforms (e. g., Shopify). Businesspeople were also alert and seized the opportunities in the new environment, especially with the increased demand for technology services
- **HealthTech and BioTech Surge:** The pandemic significantly underlined the need to revamp the healthcare sector and increase health-tech and biotech startups. Pandemic needs have led businesspersons to innovate, and different approaches to telemedicine, remote diagnostics, and vaccine production based on AI, machine learning, and data analysis exist
- **Sustainability and Green Tech:** In light of the emerging sensitive issue of climate change, technopreneurship is shifting more towards green technology. Promising business innovations across CleanTech, renewable energy, and green transportation, whether electric cars, solar power, etc., are the most active in solving environmental issues.

Technopreneurship has evolved over the years in line with the transforming technological changes that continue to affect and transform business environments. Whether it be within an industrial revolution through mechanization of industries in the 18th century, or the digital and AI revolution in today's world, technopreneurs have been pivotal in spearheading economic development and redefinition of industries and disseminating solutions to various social issues. Technopreneurship in the future is expected to integrate more technologies like AI, blockchain, and quantum computing to advance technopreneurship in different fields (Rajaram, 2023).

## 4.3 TECHNOPRENEURSHIP IN VARIOUS SECTORS

Technopreneurship is the integration of technology and entrepreneurship, which are significant in shifting the paradigms of different industries and fields through innovation, productivity, and value creation.

### 4.3.1 Information Technology (IT) and Software Development (Sharma, 2010)

- **Innovation in Software Solutions:** IT technopreneurs design and launch niche software solutions that address business and consumer issues. For instance, cloud computing services, cybersecurity solutions, and software as a service platform are products of technopreneurial businesses
- **Automation and Artificial Intelligence (AI):** AI-based startups diversify in capabilities, including learning, analytics, and robotics, to advance smart systems and perform tasks more efficiently.

### 4.3.2 Healthcare and Biotech (Paramasivan and Selladurai, 2017)

- **Telemedicine and Health Tech:** Technopreneurship facilitates the creation of telemedicine solutions, remote diagnostics, wearable health tracking devices, and other custom applications for healthcare needs. These innovations entail making healthcare more efficient and accessible to the people
- **Medical Devices and Biotechnology:** Technopreneurs in biotech assist in developing innovative equipment in medicine, diagnostics, and biopharmaceuticals. Such devices include those with tracking health features, such as wearing gadgets and AI diagnostic tools
- **Health Data Management:** While adopting big data and analytics, these technopreneurs are designing systems for managing patient data to enhance health using predictive analytics.

### 4.3.3 Education and E-Learning

- **EdTech Platforms:** Technopreneurs implement e-learning systems that deliver education content, classes, and training through the Web. Some examples of technology-aided teaching include MOOCs, tutoring, and games-based learning solutions

- **AI in Education:** Artificial intelligence in learning is helping startups develop learning solutions where learning paths are customized to the learners, where they excel through adaptive learning, and where effective student performance is enabled through intelligent tutoring systems
- **Digital Credentialing and Certifications:** New credentialing systems using blockchain are being implemented to impart secure and verifiable certifications for online learning, thus increasing the authenticity of web-based education (Hidayat *et al.*, 2019).

### 4.3.4 Manufacturing and Industry 4.0

- **Smart Factories and IoT:** Technopreneurship plays a pivotal role in defining Industry 4.0, where IoT, sensors, and big data analytics are adopted in smart manufacturing, smart production line management, and equipment monitoring and control
- **Additive Manufacturing (3D Printing):** Regarding technology innovation, various professionals within the sphere of entrepreneurship are employing 3D printing innovation to support different objectives, which include rapid prototyping and effective and efficient production, as well as customized manufacturing
- **Automation and Robotics:** Technopreneurs improve sectors like automotive, electronics, and aerospace industries with innovative robotics and automation technologies by increasing accuracy, minimizing mistakes, and increasing production rates (Zein, Ghalih and Pebriana, 2021).

### 4.3.5 Agriculture (Agritech)

- **Precision Agriculture:** Technopreneurs are developing products such as drones, sensors, and satellite images to help farmers check on the status of their crops, the amount of water to use for irrigation, and the best method to get high yields by embracing data farming
- **Agritech Platforms:** Selling agricultural produce through online markets enables farmers to eliminate intermediaries and get better prices for their produce. Blockchain technology is also applied to transparent supply chains
- **Sustainable Farming:** Technological solutions in biotechnology, like genetically modified crops and vertical farming, are focused on the sustainable practices of agriculture as a response to world hunger (Sahu *et al.*, 2018).

### 4.3.6 Energy and CleanTech

- **Renewable Energy Solutions:** Technopreneurs lead the development of solar, wind, and other renewable energy technologies. These innovations make us less reliant on fossil fuels and ensure that energy is cheap and within reach
- **Energy Storage and Smart Grids:** The young firms are investing in enhanced battery systems and smart grid technology that can improve energy distribution and minimize energy loss

- **Sustainable Technology:** Technopreneurs in the CleanTech sector are creating technologies to mitigate carbon emissions in products like electric vehicles, energy-related appliances, and green structures (Wong, 2022).

### 4.3.7 Financial Services (FinTech)

- **Digital Payment Solutions:** Technopreneurship has brought about a massive shift in cashless transactions, mobile wallets, payment gateways, and cryptocurrencies. These innovations enhance the safety, speed, and convenience of managing transactions
- **Blockchain and Cryptocurrencies:** Technopreneurs can use blockchain technology to develop the decentralized financial networks, guaranteeing a high level of transaction transparency and security. Cryptocurrencies and other financial products that use blockchain technology, such as smart contracts, are becoming popular
- **Insurtech and Wealth Management:** In the insurance and wealth management segments, fintechs are using artificial intelligence to create individual financial plans, automated investment advisory, and instant insurance solutions (Iskandar *et al.*, 2022).

### 4.3.8 Retail and E-Commerce

- **Online Marketplaces:** It is important to note that e-commerce giants such as Amazon, Alibaba, and Flipkart are credited to technopreneurs who have altered their business practices through consumption levels by providing convenience, variety, and competitive prices
- **AI and Data Analytics in Retail:** Machine learning-based recommender systems, automated inventory tracking systems, and data analytics enhance retail enterprise customer experience and organizational performance
- **Omni-Channel Retailing:** Thus, Technopreneurship allows physical and virtual store designs depending on the adopted solutions (Tanvir and Pearl, 2021).

### 4.3.9 Transportation and Mobility

- **Ride-Sharing Platforms:** In transportation, companies such as Uber and Lyft exemplify Technopreneurship since they use technology to connect passengers and drivers effectively
- **Electric Vehicles (EVs):** Many technopreneurs in the automotive industry are now opting for EVs to adopt eco-friendly and energy-efficient Technology. Some examples include manufacturers of high-end electric vehicles, such as Tesla
- **Autonomous Vehicles and Drones:** Self-driving vehicles for passengers, personal delivery services, and drones are being developed. These technologies will completely transform the logistics and transportation of goods and people to make it safer and more efficient (Torre-Bastida *et al.*, 2018).

### 4.3.10 Entertainment and Media

- **Streaming Services:** Technopreneurs have impacted entertainment through innovations such as Netflix, YouTube, and Spotify, which allow consumers to access media content at their convenience
- **Augmented Reality (AR) and Virtual Reality (VR):** Startups exploit these technologies to deliver gaming, movie, and advertising experiences
- **Content Creation and Distribution:** Social media and digital marketing technologies such as TikTok and Instagram offer new equally monetized content through the dawn of Technopreneurship (Kamarudin and Sajilan, 2013).

### 4.3.11 Real Estate (Proptech)

- **Smart Buildings and IoT:** Technopreneurs are creating smart building technology incorporating IoT for power, security, and maintenance in business and private environments
- **Property Management Platforms:** Some of the technologies in the real estate field involve using apps that help with property management, selling or renting houses and apartments, and smart contracts based on the blockchain
- **Virtual Reality Tours:** Through immersion technologies, real estate is marketed and sold by providing constructed virtual reality tours to the anticipated users.

Technopreneurship is an innovative way that fuels change throughout various sectors, including business, where technology is combined with business. It increases economic growth, generates employment, and helps solve significant world issues, such as environmental protection and populations' access to healthcare and education. In the future, technopreneurs will expand and adapt to different fields, and as technology advances, innovative ideas and possibilities in a given field will probably be brought about by technopreneurs (Nafiisa, Jaswadi and Djajanto, 2024).

## 4.4 BENEFITS AND LIMITATIONS OF TECHNOPRENEURSHIP

### 4.4.1 Advantages of Technopreneurship:

- **Innovation and Disruption:** A technopreneur can create new opportunities and change established industries by introducing new products, services, and technologies. Some examples of such technologies are the ride-sharing applications Uber and Lyft and commerce platforms Amazon and Alibaba, which have disrupted transportation and retail industries
- **Economic Growth:** Technopreneurs play a major role in constructing economic status, generating employment opportunities for people, and improving economic competitiveness. Most start-ups and industries with a technological base comprise most countries' gross domestic product

- **Scalability and Global Reach:** New technology-based economy businesses have high scalability. It can easily move from a local market to a world market with a little extra overhead. An example of how existing industries can achieve economies of scale is by considering the capacity of digital platforms to reach millions of clients globally at a relatively low cost
- **Increased Efficiency and Productivity:** Technopreneurship enhances innovations that increase the effectiveness in accomplishing activities and streamlining processes across organizations. Another application of AI, machine learning, and other technologies, such as IoT, is increased productivity by having them perform high error rates and time-consuming tasks
- **Social and Environmental Impact:** Several technopreneurial initiatives have been created that address social and environmental issues, including power through solar and wind energies, agritech, and medicine through telemedicine diagnosis and diagnostic devices. They have a positive social aspect and achieve this by solving problems that conventional business organizations do not solve
- **Accessibility and Democratization:** Technopreneurship frequently facilitates easier access to services like education, finance, and health by providing affordable solutions to target consumers who have not been served adequately. For instance, mobile payment and online classes have made essential services available in rural regions
- **Flexibility and Adaptability:** Tech-based businesses are more adaptable to the changing market demand than the conventional economy since they can change their business models in response to evolving market conditions. This makes them ready to adapt within a short period to a competitive world that is rapidly changing.

### 4.4.2 Disadvantages of Technopreneurship

- **High Risk and Uncertainty:** Consequently, the prospect of technopreneurship entails considerable risk or uncertainty, particularly during the initial years of startup operations. New ventures will likely experience conditional ambiguity, including uncertainty about customer demand, competitors, and technology risks. Technology ventures suffer high mortality rates because they become non-viable or financially ill-equipped
- **Dependence on Technology:** Over-reliance on technology poses significant risks for various incidents such as cyber threats, loss of data, or system breakdown. Also, the ever-evolving technology presents a problem in that technopreneurs are compelled to continuously update their solutions, which can be expensive and labor-intensive
- **Capital Intensive:** Almost all technopreneurial ventures, especially those operating in sectors such as hardware, AI, biotechnology, and renewable energy sources, often involve substantial capital costs in technology and talent acquisition. It can be difficult to convince investors to fund startups, and many businesses face cash flow problems
- **Skill and Talent Shortage:** Technopreneurship needs key talents to drive its success in software development, artificial intelligence, data science, and

engineering. However, there is often a scarcity of such talent, which can hinder the rate at which products are developed, or even innovations are made

- **Rapid Market Saturation:** Technology space is very congested, and any idea, product, or service developed in this area can easily be emulated or enhanced by rivals. This results in increased market competition and product life cycle shortening, which may pressure the technopreneurs to sustain and expand continuously
- **Regulatory and Legal Challenges:** Technopreneurs experience various legal challenges, including licensing issues, particularly in sectors heavily dependent on regulation, such as fintech, healthcare, and biotechnology. Legal concerns such as data privacy laws, intellectual property rights, and compliance issues are often constraints that may limit the growth of tech startups
- **Work-Life Imbalance:** One of the effects of Technopreneurship is that the business owners spend most of their time working to meet the set objectives of the business, hiring employees, or handling challenges. They suffer from burnout and stress, influencing their career and private lives
- **Technological Obsolescence:** Technological breakthroughs have the potential to make previously relevant products and services outdated quickly. Innovativeness is another essential aspect of the technopreneurial success, meaning there must be constant research and development investment to adapt to the changing trends. Failure of this can lead to a company becoming irrelevant in its market.

Technopreneurship presents an avenue for change, economic development, and social welfare but not without potential pitfalls, such as high capital investment demands, market volatility, and technological issues. To the entrepreneurs willing to venture into this area, it is important to note that while these things are an advantage, there is also the need to adopt sound management practices, strategic planning, and flexibility in project implementation.

## 4.5 PERFORMANCE ANALYSIS OF TECHNOPRENEURSHIP

Bibliometric analysis is a research methodology that applies quantitative methods to study the trends and compile, share, and systematically organize data based on published literature in a given field of study (Singh *et al.*, 2023). Different analysis indicators like citation index, publication frequency, and author activity are all possible through bibliometric analysis and support the researcher in identifying the ontogeny of a particular scientific field, prominent works, and trends. This method is particularly important in evaluating the success of scientific production and the relevance of the obtained results in quickly evolving fields (Jain *et al.*, 2023). The reason for carrying out a bibliometric analysis is to conduct a systematic mapping of the known research arena, determine path-breaking articles and authors, and discover new paradigm shift directions in a particular field. This, in a way, can help give an overview of the current area of specialization, its historical development, and main contributions to the field and assist in identifying the research environments. This can help determine the coming research agendas and aid policy formulation and academic decisions. To achieve the desired objectives in this study, a method of bibliometrics analysis was performed on the Scopus database, one of the researchers biggest and most popular databases (Sharma, Sakhuja and Nijjer, 2022). The selected search string was (TITLE-ABS-KEY(("Technopreneurs") OR ("Technopreneurship"))), which was expected to identify articles that focus on Technopreneurship. After fulfilling the initial search criteria, the number of articles obtained was 180.

## 4.6 YEAR WISE ANALYSIS

In the research paper, the author has provided a wise distribution of data from the year 2000 to the year 2024. Hence, comparing the publications or their occurrence in different countries shows that the numbers are disparate and exhibit different trends over the years. Afterward, there were few filings until the early 2000s, with a marginal increase from 2 in 2000 to 3 in 2001. In subsequent years, they recorded a lower figure in 2002 (2002 = 1), though it rose in 2004 (2004 = 5). Some fluctuations can be seen; they have some years with five deaths and then one or two deaths in other years, such as in 2004 and five deaths in 2006, followed by one death in 2005 and one death in 2008. Since 2017, a significant increase has become observable, distinguishing the current situation from the previous years when the stated figure was relatively low. It is noted that there is a fair amount of activity or attention to the subject matter in these years: 2017 records 14 occurrences, and 2018 records nine occurrences; however, interest in the subject matter increases in 2019 with 14 occurrences and escalates notably in 2020 with 18 occurrences. The years 2021 through 2024 show an increase in the trend, and 2023 records the highest number of occurrences, with a total of 29. Even though in 2024, it went back down to fifteen, it is still considered much higher than in the early years. These escalations reveal that more attention or focus has recently been given to the topic under study, especially between 2020 and 2023. Such outcomes indicate increased enrollment and, consequently, interest in the subject matter, which may stem from new trends, Technology, or research financing in the later years of the timelineyear wise analysis represented in Figure 4.1.

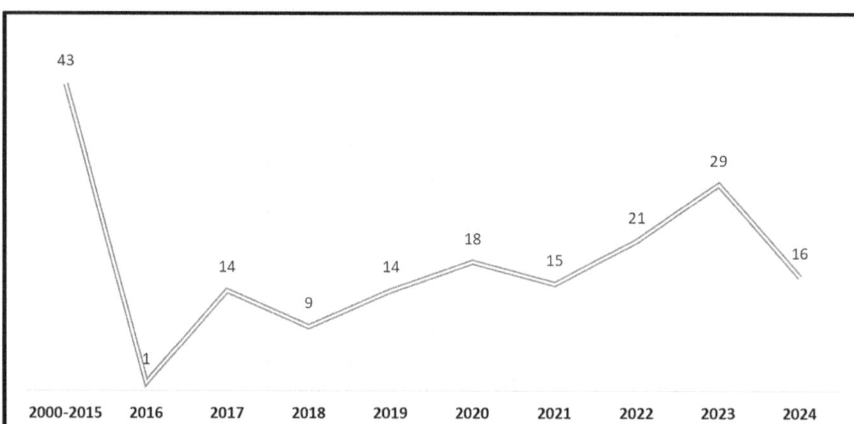


**FIGURE 4.1** Year Wise Analysis.

## 4.7 LEADING SOURCES OF PUBLICATION

In this case, the analysis of the research contribution goes beyond focusing on the number of papers published in a given period; it also covers the dissemination and relevance of these papers across different journals and conference proceedings, as represented in Figure 4.2. Other venues include the AIP Conference Proceedings, where five publications have been published and are popular among physicists and engineers, providing a strict peer review process and listed in Scopus and Web of Science. The *International Journal of Technoentrepreneurship* has had four publications and falls under the category of technoentrepreneurship, which deals with the use of technology in entrepreneurship; it is also indexed in esteemed databases, and its impact factor is average within the specific field. The IOP Conference Series: The Earth and Environmental Science journal has published four papers, the aim of which is to focus on identifying peer-reviewed research within the environmental sciences to retain high visibility even with a lower impact factor that is common in conference series. Industrial Engineering & Operations Management is another four-publication conference acknowledged for its significance in industrial engineering and operations management, highlighting the applicability of its results. Yet, impact factors are not usually determined for meetings. Finally, ACM International has three publications and appears to be the most popular computing and information technology publication. It has high citation scores but without a direct score factor. In summary, the study highlights a broad spectrum of well-indexed publication outlets in various disciplines, confirming their role in sharing new knowledge and promoting scholarly connections while suggesting that conference proceedings, which are generally posted at significantly lower journal-ranking levels, may remain a valuable communication method.

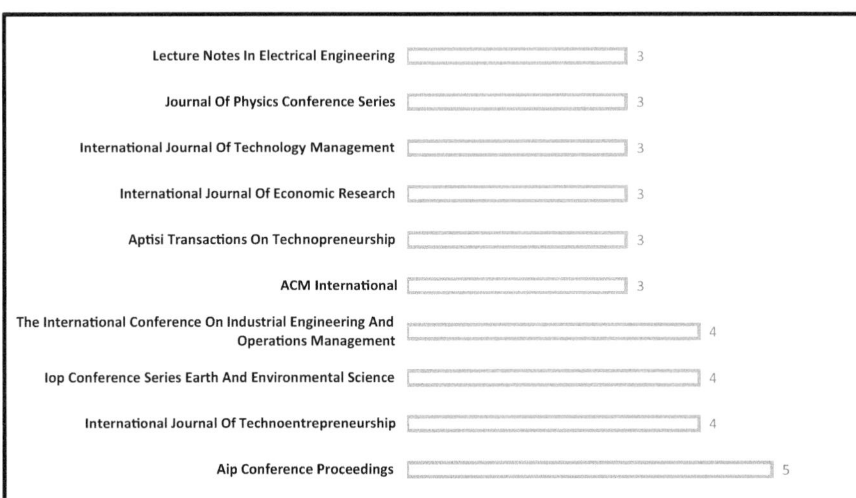


**FIGURE 4.2** Leading Source Analysis.

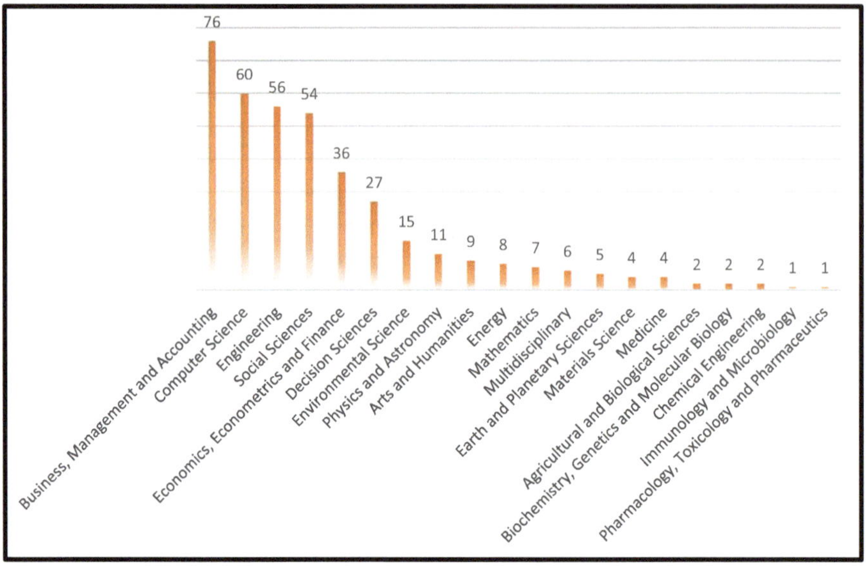


**FIGURE 4.3** Subject Area Analysis.

## 4.8 SUBJECT AREA ANALYSIS

The distribution of the analyzed publications by the disciplines shows that the research on cloud computing and IoT is interdisciplinary, as depicted in Figure 4.3. The data analysis reveals that most articles are published in business, management, and accounting, with 76 articles; computer science, 60, and engineering, 56. These fields remain the most prolific in their contribution to academic literature owing to their relevance in championing innovation, technology, and corporate responsibility. social sciences (54) also remains relevant since it focuses on societal behavioral patterns and trends. Finance (36) infers the organizations' financial systems and their modeling, and economics and econometrics (36) suggest its emphasis on the systems of economy and econometric modeling for financial organizations. Though smaller in comparison, environmental science (15) also indicates a shift toward greater awareness of environmental and climate concerns. This distribution indicates a central concern with technology and social concerns in the recent studies.

## 4.9 AUTHOR CO-AUTHORSHIP NETWORK ANALYSIS

Sutopo has published seven documents and has been cited 11 times, which makes him the most productive author. Although his work has been cited comparatively less, it indicates a steady productivity rate in the research world. Hisjam has authored and co-authored five documents with five citations, which signifies his balanced participation in writing the document and being cited. Malhotra, Harpreet Kaur has published five documents. Still, these documents received only 16 citations, while Ismail, Kamariah,

**FIGURE 4.4** Author Co-authorship Network Analysis.

with fewer published documents (4 documents) with 20 citations, shows that her work is more impactful or visible within the academic fraternity. All three authors, Herawati, Susi, and Hidayat, Hendra, have contributed three documents each. Still, the documents are well cited and have been cited 50 times each, demonstrating their societal impacts.

In conclusion, it is observed that the number of publications for these authors differs from one another. However, the reported citation demonstrates the researchers' importance and contribution to their respective fields. Authors' network analysis is represented in Figure 4.4.

## 4.10 COUNTRY WISE ANALYSIS

The results above show the top countries that have contributed to the published articles by researchers through various sources, as depicted in Figure 4.5. Indonesia tops the list with 57 articles, implying the country's increasing contribution to knowledge production in various areas, including education, technology, and economics. Malaysia comes next with 35 articles, illustrating its growing research activities and increasing emphasis on higher learning and research and development funding. It should be noted that Indian scholars have made significant contributions to the scientific literature, a record totaling 15 contributions was achieved. Singapore (14 publications) continues to be very active in producing high-impact research articles, particularly in technology and business. Australia (8 publications), with its strong academic and research environment, evidently contributes important work in various fields. This geographical distribution shows how researchers worldwide are involved in academic research with more focus on Southeast Asia.

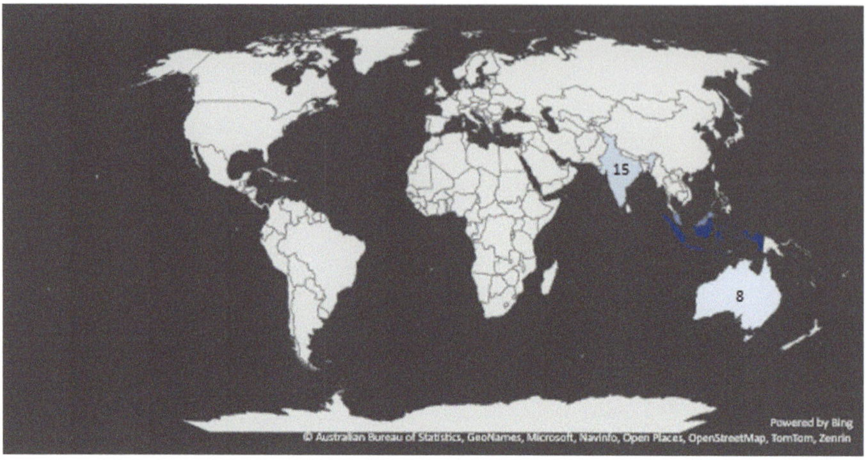


**FIGURE 4.5** Network Analysis of Countries.

**FIGURE 4.6** Word Cloud Keyword Analysis.

## 4.11 KEYWORD ANALYSIS

The keyword analysis in the bibliometric analysis provides insights into the trends and focal points in the research domain. The term "technopreneurship" is the most frequent, with 75 suggesting continued research interest in combining technology and entrepreneurship, possibly examining the influence of technology in business ventures. 'Entrepreneurship' (29) and 'innovation' (16) are also represented strongly, suggesting a strong research focus on the role of entrepreneurship and innovative practices. The keywords "students" (16) and "engineering education" (11) hint at the development of a culture of innovation and entrepreneurship among the young generation with the use of education. There is a focus on "technology transfer" (11), which shows the concern solicited regarding how technology spreads from the research stage to the commercial stage. Other key terms include 'commerce' (9), 'information technology' (8), and 'technopreneur.' (14), 'Singapore' (8), and all of them point to the fact that business is intertwined with technology. In conclusion, the keyword analysis indicates a rich research area focused on technology and innovation in learning and commerce. The word cloud of keywords is illustrated in Figure 4.6.

## 4.12 CONCLUSION AND FUTURE SCOPE

This research has discussed the development of technopreneurship and its importance in advancing Technology and boosting the economy. Technopreneurship has played a crucial role in reinventing industries through technology and innovation, particularly in healthcare, education, energy, and agriculture. Bibliometric analysis has increased research interest in technopreneurship among researchers from different countries. The study has also focused on the responsiveness of technopreneurs to global problems such as global warming, resource depletion, and famine by developing technology-based solutions.

- **Artificial Intelligence (AI) and Machine Learning:** These technologies will remain prominent and shape innovations in different sectors, including care, finance, and others
- **Blockchain and Cryptocurrencies:** The adoption of blockchain in sectors such as finance and supply chain is anticipated to expand, as it provides a secure decentralized platform for points of sale and storage of information
- **Quantum Computing:** This new field is expected to offer immense computational resources, effectively changing facets of life like drug discovery, material science, and cybersecurity
- **Sustainability and Green Tech:** Due to rising environmental consciousness, technopreneurship will involve developing green technologies in energy and agriculture
- **Internet of Things (IoT):** The further incorporation of IoT technology into different fields will progress smart homes, cities, and industries, increasing the system interconnectivity and automation.

Technopreneurship shall continue to play a critical role in the co-creation of the future, the contribution representing innovations that solve worldwide problems and generate economic returns.

## REFERENCES

Ab Rahman, W.N.W. *et al.* (2024) 'Model for responsive agriculture hub via e-commerce to sustain food security', *International Journal of Advanced Computer Science & Applications*, 15(5).

Adeoti, J.O. (2019) 'Technopreneurship and national development: Powering Businesses and the Economy with New Technologies', *Nigerian Institute of Social and Economic Research* [Preprint].

Choi, L.K., Iftitah, N. and Angela, P. (2024) 'Developing technopreneur skills to face future challenges', *IAIC Transactions on Sustainable Digital Innovation (ITSDI)*, 5(2), pp. 127–135.

Dada, A. (2021) 'Digital entrepreneurial behaviour of millennials in the fourth industrial revolution in southwest Nigeria: Digital entrepreneurial behaviour of millennials', *African Journal of Science Policy and Innovation Management*, 2(1&2), pp. 41–60.

Diantoro, K. *et al.* (2023) 'Implications of distributed energy development in blockchain-based institutional environment', *Aptisi Transactions on Technopreneurship (ATT)*, 5(2sp), pp. 209–220.

Dolatabadi, R.V. and Meigounpoory, M.R. (2013) 'Effective determinants of corporate nano-technopreneurship process in active technological knowledge base firms', *International Journal of Academic Research in Economics and Management Sciences*, 2(5), p. 137.

Hidayat, H. *et al.* (2019) 'The contribution of technopreneurship scientific learning and learning readiness towards the entrepreneurship learning outcomes in higher vocational education', *Jurnal Pendidikan Vokasi*, 9(1), pp. 21–32.

Iskandar, R.D.M. *et al.* (2022) 'Indonesian islamic banking fintech model strategy: Anp method', *Aptisi Transactions on Technopreneurship (ATT)*, 4(2), pp. 142–152.

Jain, S. *et al.* (2023) 'Blockchain and cryptocurrency: A bibliometric analysis', *Journal of Advanced Computational Intelligence and Intelligent Informatics*, 27(5), pp. 822–836.

Kamarudin, H.S. and Sajilan, S. (2013) 'Critical success factors of technopreneurship in the creative industries: A study of animation ventures', *Review of Integrative Business and Economics Research*, 2(1), p. 1.

Milton-Smith, J. (2003) 'Developing technopreneurs in Asia: A strategic management research perspective'. Working Paper Series 2005: no. 2003:1, Curtin University of Technology, School of Management. https://espace.curtin.edu.au/handle/20.500.11937/18668

Nafiisa, B.L., Jaswadi, J. and Djajanto, L. (2024) 'PIECES framework for information system needs analysis using website-based Appsheet: A technopreneurship digitalization solution in Indonesia', *East Asian Journal of Multidisciplinary Research*, 3(7).

Ndagi, A. and Salihu, A.A. (2019) 'Technology entrepreneurship and fourth industrial revolution era', *LAPAI International Journal of Management and Social Sciences*, 11(1), pp. 142–153.

Olusegun, O.O., Akpoviroro, K.S. and Adebowale, O.M. (2019) 'Impact of technopreneurship on business performance', *Journal of Economics and Management Research*, 8, pp. 55–71.

Paramasivan, C. and Selladurai, M. (2017) 'Technopreneurship education: Teach and train the youths', *Asian Journal of Management*, 8(4), pp. 1173–1176.

Rajaram, K. (2023) *Leading and Transforming Organizations: Navigating the Future.* Routledge.

Sahu, K.P. *et al.* (2018) *National Youth Convention on Transforming Agriculture for Agripreneurshipand Employment Opportunities: The Youth Perspective: Souvenir & Abstract.* All India Agricultural Students Association 3-11.

Sharma, C., Sakhuja, S. and Nijjer, S. (2022) 'Recent trends of green human resource management: Text mining and network analysis', *Environmental Science and Pollution Research*, pp. 1–20. Available at: https://doi.org/https://doi.org/10.1007/s11356-022-21471-9.

Sharma, R.R. (2010) 'Technopreneurship in India: Two case studies of information technology entrepreneurs', in *Handbook of Research on High-Technology Entrepreneurs.* Edward Elgar Publishing.

Shekhar, S. *et al.* (2023) 'Academic bioinformatics activities joining hands with entrepreneurial ventures: A way to go in life science research (case studies from Indian academia, start-ups and the life science sector)', *International Journal of Research, Innovation and Commercialisation*, 5(1), pp. 70–105.

Singh, S. *et al.* (2023) 'Making sense of glass ceiling: A bibliometric analysis of conceptual framework, intellectual structure and research publications', *Cogent Social Sciences*, 9(1), p. 2181508.

Torre-Bastida, A.I. *et al.* (2018) 'Big Data for transportation and mobility: Recent advances, trends and challenges', *IET Intelligent Transport Systems*, 12(8), pp. 742–755.

Tyagi, A.K., Aswathy, S.U. and Abraham, A. (2020) 'Integrating blockchain technology and artificial intelligence: Synergies perspectives challenges and research directions', *Journal of Information Assurance and Security*, 15(5), p. 1554.

Wong, P.K. (2022) 'The development of Singapores innovation and entrepreneurship ecosystem', in Engel, J. (ed.) *Clusters of Innovation in the Age of Disruption*. Edward Elgar Publishing, pp. 206–244. https://doi.org/10.4337/9781800885165

Woo-Je, C.H.O. (2024) 'The history of tourism distribution channels and future prospects in the tourism service industry', *Journal of Distribution Science*, 22(6), pp. 107–114.

Xiong, J. (2016) 'Understanding the role of information technology for sustainable development in small businesses and micro-enterprises', *Student Work*, 2910. https://digitalcommons.unomaha.edu/studentwork/2910

Zein, M., Ghalih, M. and Pebriana, R. (2021) 'Entrepreneurship and technopreneurship in Era 4.0: GO-JEK extended to Decacorn', in *Handbook of Research on Innovation and Development of E-Commerce and E-Business in ASEAN*. IGI Global, pp. 624–639.

# 5 Digitalization in Business Growth and Scaling

*Abhishek Bhattacherjee, Ajay Kumar Badhan, Raman Kumar, and Swapandeep Kaur*

## 5.1 INTRODUCTION TO DIGITALIZATION

Businesses, in the world of fast-paced growth in technology and rapidly changing consumer preferences, are under greater pressure than ever to change or be replaced. That requirement is born of digitalization, the powerful force that is reshaping economics, industries, and even commerce itself. Digitalization, often used interchangeably with terms like digital transformation or digitization, refers to digital technology integration into all areas of business operations and company functions. It involves bringing in digital platforms, tools, and methods that promote efficiency, growth, and innovation.

The progression of becoming digitally literate requires more than just new software. It demands a shift in the perspective and the mindset of implementing different methodologies in businesses. The companies must embed customer-centricity, data-driven decision-making, and agility into the core of how they operate.

### 5.1.1 Evolution of Digitalization

The origin of digitalization can be traced back to the development of computers and the internet. However, with the improvements in artificial intelligence, machine learning, cloud computing, and big data analysis, technology has significantly developed over the years, making it a powerful tool for business use. From multinational corporations to emerging startups, organizations across industries adopt digitalization as a new wave of opportunity to take advantage of and enhance efficiency with the scope for sustainable growth. Digital technologies can be used in many ways, from creating unique digital products, to how clients interact with businesses and supply chain efficiencies.

Computing and data management were redefined with the arrival of increased machine processing capabilities. The beginning of the digitization era is represented in Figure 5.1.

Early Internet: By creating the foundation for worldwide connectedness, the early internet made it possible to communicate and share information on a previously unthinkable scale. The modern networked society we live in was made possible by the development of basic networking protocols and technologies.

 DOI: 10.1201/9781003537335-5

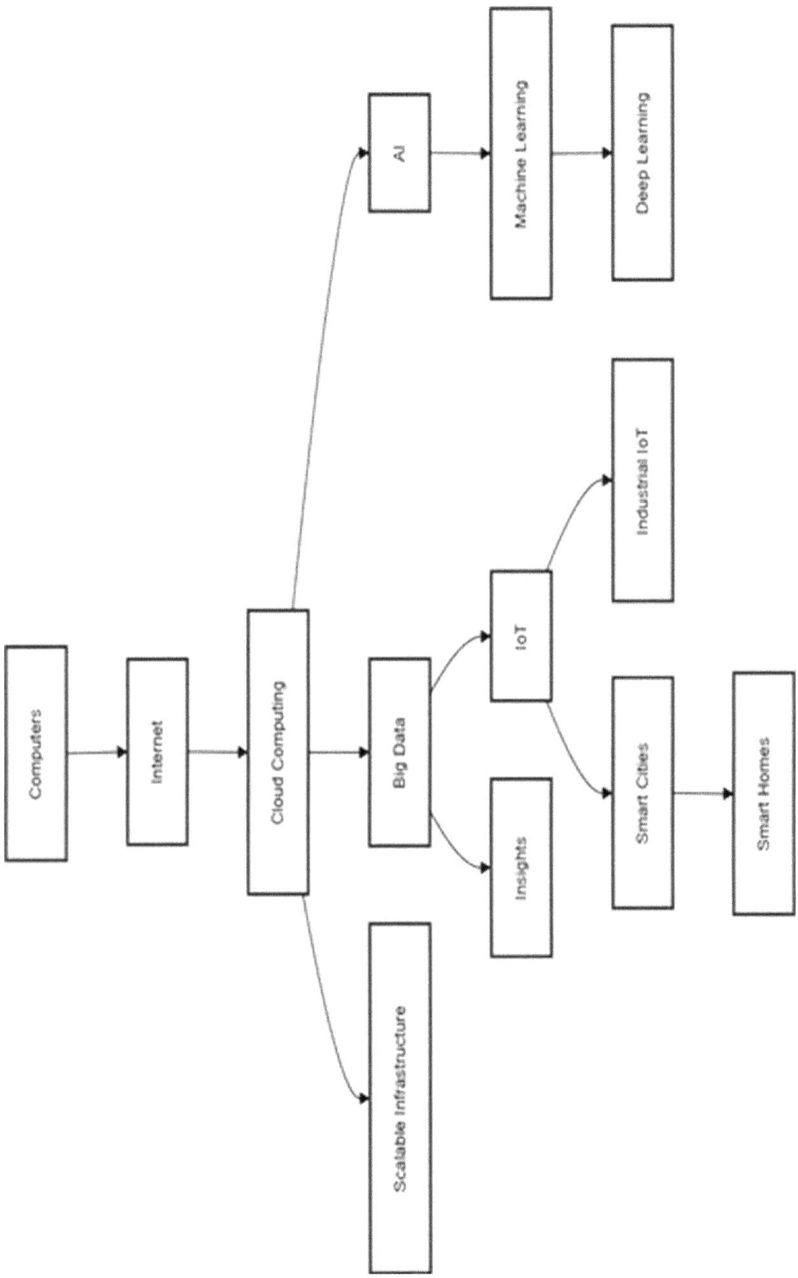


**FIGURE 5.1** Evolution of Digitalization.

Technological Advancements: As technology developed, significant innovations surfaced in fields including big data analytics, cloud computing, artificial intelligence, and the Internet of Things (IoT). These developments sparked a subsequent wave of digitalization, which changed entire industries and the way companies functioned. The diagrammatic view of it is presented in Figure 5.1.

### 5.1.2 Digitalization in the Context of Businesses

When used in a business context, digitalization is the process of incorporating digital technologies into all facets of an organization's operations, procedures, and plans in an effort to boost productivity, creativity, and competitiveness. It entails transforming conventional business procedures and adjusting to the needs of the digital age by utilizing digital tools, platforms, and methodologies in these sectors:

| | |
|---|---|
| 1. | Automation |
| 2. | Data-driven Decision-making |
| 3. | Customer Experience Enhancement |
| 4. | Agility and Flexibility |
| 5. | Digital Innovation |
| 6. | Strategic Partnerships |

### 5.1.3 Impact on Industries

A structured outline of the impact of digitalization on various industries can be seen represented in Figure 5.2.

Figure 5.2 represents numerous industries that have been significantly impacted by digitalization, which has welcomed a new period of innovation, efficiency, and connectedness. E-commerce platforms have made shopping experiences more personalized and global, while data analytics have improved pricing and inventory control in the retail industry. Fintech innovation and blockchain technology are transforming services and enhancing transparency in the finance sector. Wearable technology, electronic health records, and telemedicine all improve patient care and accessibility in the healthcare industry as shown in Figure 5.2. Automation and predictive maintenance boost productivity in industry, while ride-sharing and driverless cars in transportation provide safer, more effective mobility options. Streaming services, social networking platforms, and gaming technologies are the lifeblood of media and entertainment, revolutionizing the way people engage with and consume content. Digitalization brings issues such as worker adaptation and cybersecurity, but it also promotes agility, competitiveness, and customer-centricity across industries.

Numerous disruptive breakthroughs have been made in several fields, including marketing and accounting. With a projected value of USD 880.28 billion in 2023, the global market for digital transformation is predicted to expand at a compound annual growth rate (CAGR) of 27.6% between 2024 and 2030. In the previous few decades, troublemaking technology has also become a part of our daily life.

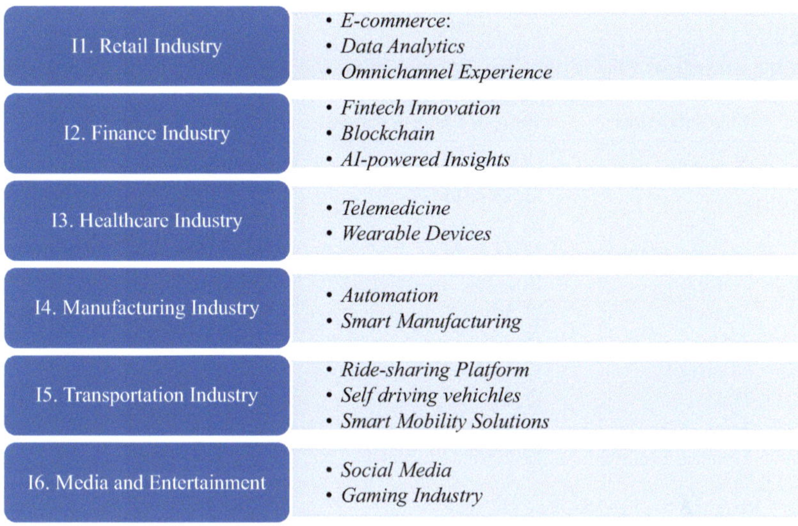


**FIGURE 5.2** Impact on Industries.

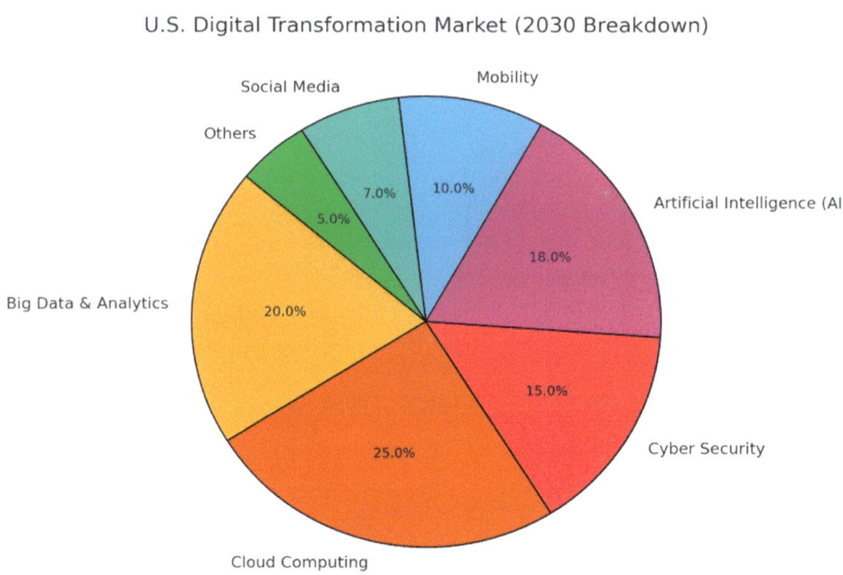


**FIGURE 5.3** Digital Transformation Statistics of United States. (Retrieve Kraus et al. 2021.)

Our habits match technological advancements like the internet and cell phones and its services perfectly. A disruptive technology is defined as an invention that profoundly alters consumer, industry, or business behavior. The study progress is shown in Figure 5.3.

**TABLE 5.1**
**Top Organization Digitalization Approaches**

| S. No. | Organization | Type of Organization | Digitalization Approach |
|---|---|---|---|
| 1. | Amazon | I1, I4 | For doing Digital Transformation / Marketing / Communication / Transactions / Healthcare / Education of different processes these organizations are using AI, Cloud Computing, Big Data & IoT. |
| 2. | Apple | I2, I3 & I4 | |
| 3. | Tencent | I6 | |
| 4. | Google/Alphabet | I2, I3 & I4 | |
| 5. | Netflix | I6 | |
| 6. | SpaceX | I2, I4 | |
| 7. | Tesla | I5 | |
| 8. | Microsoft | I2, I3 & I4 | |
| 9. | IBM | I2, I4 | |
| 10. | Intel | I2, I4 | |
| 11. | General Electric | I5 | |

Source: Retrieved from Vasilev et al. (2020).

## 5.2 THE ROLE OF DIGITALIZATION IN BUSINESS GROWTH

In today's digital economy, digitalization plays a critical role in business growth by acting as a catalyst for efficiency, innovation, and competitiveness. Businesses may create new opportunities, improve customer experiences, and expedite procedures by incorporating digital technologies into every part of their operations. Through digitalization, businesses can increase market reach, generate new revenue sources, and connect with a larger audience via online platforms. Businesses can use data analytics to obtain insightful information about market trends, client preferences, and operational performance. This information can then be used to inform data-driven decision-making and focused marketing campaigns. Automation of repetitive operations and workflows boosts output, lowers expenses, and frees up staff time for high-value work.

As per Lichtenthaler (2018), there are a total of 11 top inventive organizations in the worldmaking the best use of digitalization. These are listed in Table 5.1 with more than 3 ratings.

### 5.2.1 Advantages of Digital Technologies

Businesses can benefit from a wide range of digital technology benefits, such as increased efficiency, enhanced customer experience, and a wider market reach. By leveraging digital platforms and focused marketing strategies, companies may expand their client base globally and penetrate untapped regions. Process simplification and automation maximize operational effectiveness, reducing time and costs and quickening time-to-market. Furthermore, data analytics offer insightful information that is helpful for strategic planning and well-informed decision-making. Furthermore, through self-service options and multi-channel communication, digital technologies provide tailored customer experiences that strengthen relationships and

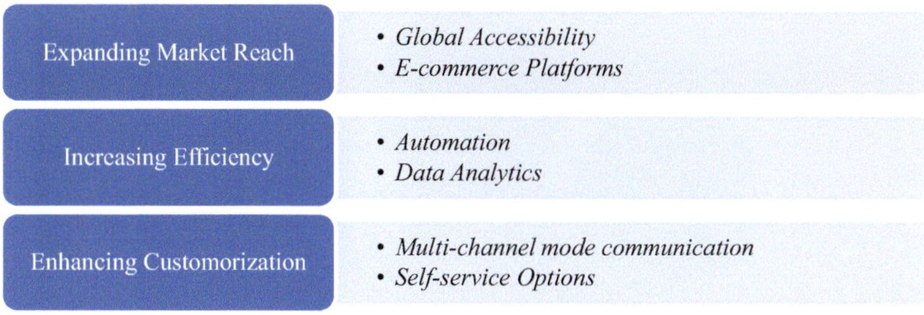


**FIGURE 5.4** Advantages of Digitalization.

encourage loyalty. All things considered, companies must embrace digital transformation if they want to prosper in the current digital-first economy and stay competitive. The diagrammatic view is represented in Figure 5.4.

For example, by utilizing digitization to offer a flawless online buying experience, Amazon transformed the retail sector. Amazon has diversified into other industries, including digital content (Amazon Prime Video) and cloud computing (Amazon Web Services), and it has broadened its market reach globally through its e-commerce platform, personalized recommendations, and effective logistical network.

## 5.3 KEY COMPONENTS OF DIGITAL TRANSFORMATION

In short, one may contend that there are certain cornerstones of effective initiatives in digital transformation through streamlined processes that come with efficiency, adaptability, firms being agile and flexible toward changes in technology or conditions within the market. Business models are re-examined using technology to create a digital strategy to create new revenue streams and value propositions. Personalized experiences and seamless interactions sustain the customer at the heart of the operation, which leads to consumer delight and loyalty. A technology approach continues to stand at the core of enabling digital transformation, helping businesses harness resources such as automation, cloud computing, and data analytics to spearhead innovation and competitiveness. All these key factors have a strong foundation in allowing organizations to manoeuvre successfully and profit in the digital age (represented in Figure 5.5).

The various elements depicted in Figure 5.5 have a significant impact on the expansion and scalability of businesses:

- **Process:** An automated streamlined process enables greater efficiency, cost-effectiveness, and saves time between order and delivery. It helps businesses in scaling their operations practically without sacrificing quality while increasing demand by automating the mechanicalistic of routine processes and friction points
- **Business Models:** This makes use of unconventional business models to help firms generate additional revenue, tap into the unexplored markets, and differentiate themselves from competitors. It may also present opportunities for long-term success by adapting to new trends and market changes

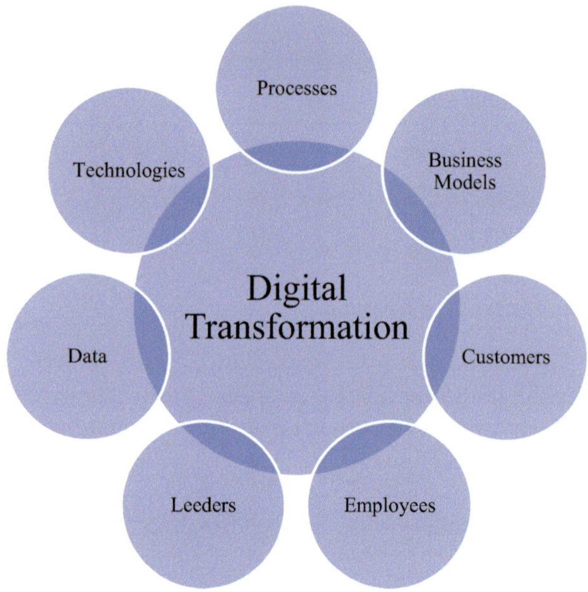


**FIGURE 5.5** Digital Transformation Components.

- **Customers:** The ultimate intention of customer-centricity is to generate more revenue, improve profitability, and retain customers. Businesses will get better insights into their customers' needs and preferences through data analytics and a tailored experience, thus deepening the relationship and encouraging repeat business
- **Employees:** Employees need to be empowered and engaged for growth and scalability in a business. Training, developing skills, and opportunities for career advancement provide a motivated and effective workforce, which ultimately leads to innovations and teamwork, ensures good customer experience
- **Leaders:** Quality Leadership: Leaders provide vision, guidelines, and support for the digital transformation. Strong leaders who possess a growth mindset build scalable environments. This mindset fosters innovation, agility, and continuous improvement. This will likely motivate employees towards the change, innovation, and transferring with the changing trends of the marketplace
- **Data:** Data-driven insights increase operational efficiency, enhance marketing capabilities, and inform strategic decisions. Technologies in data analytics and business intelligence help an organization identify trends, patterns, and growth opportunities, enabling more informed and proactive decisions
- **Technologies:** Digital technologies like cloud computing, artificial intelligence, and data analytics can make businesses more innovative, productive, and competitive. Investment in digital tools and technology infrastructure will raise efficiency, accelerate innovations, and unlock new growth opportunities.

## 5.4 CHALLENGES WITH BARRIERS IN DIGITALIZATION

Even though becoming digital can be advantageous for any organization, there are a number of obstacles that businesses must overcome. Among the most significant obstacles are:

a. **Legacy Systems and Infrastructure:** This has meant that many organizations have continued using outdated infrastructure or systems that would not necessarily be compatible with current digital technology. Overhauling, upgrading, or replacing these systems takes time and is always costly, and depletes critical resources for the operation (Anthony Jnr, 2024)
b. **Data Security and Privacy Concerns:** There is an increased risk of cybersecurity attacks and violations against data privacy when it comes to business processes connected digitally. For this reason, effective risk management practice is good to combine with firm cybersecurity protocols so as to safeguard sensitive data according to compliance with legal aspects like CCPA or GDPR regulation (Regulations and Suleh-yusuf, 2020)
c. **Skills Gap and Talent Shortage:** It would require very skilled professional competencies at different levels, including working knowledge of cybersecurity, cloud computing, AI, and big data analytics. Recruiting, onboarding, and retaining the best talent can become a strenuous exercise (Berger and Benedikt Frey, 2016)
d. **Resistance to Change:** Individuals using the older processes and systems would not ready to switch with the new when digital transformation begins. Such actions limit organizational change management. Leadership support, an effective plan of international communications, and proven change management tools are essential to overcome resistance to change in the process of change management (Bröckling et al., 2021)
e. **Integration and Interoperability:** Getting much back from digitalization tends to require that platform applications and data sources can work together. Combining multiple digital technologies and systems is complicated (Pan et al., 2021)
f. **Cost and Return on Investment (ROI):** To complete digital transformation, businesses need significant investment in software development, talent acquisition, and technical infrastructure. The return on investment is hard to measure, and actual financial value can be harder to demonstrate, at least in short term (Mardiani et al., 2023)
g. **Regulatory and Compliance Issues:** Digitalization initiatives must deal with complex regulatory frameworks and compliance standards, particularly in well-regulated sectors like finance, healthcare, or telecommunication. Compliance with industry norms and laws add another layer of difficulty to digital transformation projects (Hidayat and Defitri 2024)
h. **Digital Divide and Accessibility:** In marginalized parts of society, where people may have limited access to internet connectivity and digital infrastructure, issues such as socioeconomic inequality and problems caused by the digital divide may hinder the implementation of digital technology. Inclusive digitalization initiatives must guarantee fair access to digital resources and remove obstacles to digital literacy (Colding et al. 2024).

### 5.4.1 Strategies for Overcoming These Challenges

Organizations may counteract the hurdles and challenges of digitalization using the ways mentioned in Table 5.2:

**TABLE 5.2**
**List of challenges and their respective ways to overcome it**

| S. No | Challenges | Ways to overcome |
|---|---|---|
| 1. | Legacy Systems and Infrastructure | • Assessment and Prioritization<br>• Incremental Modernization<br>• Migration to Cloud-Based Solutions<br>• Integration and Interoperability<br>• Modularization and Microservices Architecture<br>• Invest in Modernization Tools and Technologies<br>• Change Management and Training<br>• Collaboration with Vendors and Partners (Pau et al. 2022) |
| 2. | Data Security and Privacy Concerns | • Comprehensive Risk Assessment<br>• Data Classification and Governance<br>• Encryption and Access Control<br>• Regular Security Audits and Monitoring<br>• Employee Training and Awareness<br>• Vendor Management and Due Diligence<br>• Data Minimization and Retention Policies<br>• Incident Response and Breach Management<br>• Regulatory Compliance and Accountability (Ali et al. 2024) |
| 3. | Skills Gap and Talent Shortage | • Invest in Employee Training and Development<br>• Collaborate with Educational Institutions<br>• Promote Cross-Functional Collaboration<br>• Offer Good Compensation and Benefits<br>• Deliver Opportunities for Career Advancement<br>• Embrace Diversity-Inclusion<br>• Explore Flexible Effort Arrangements<br>• Tap into External Talent Pools<br>• Promote Employer Branding and Reputation (Sundermeier et al. 2020) |
| 4. | Resistance to Change | • Communicate the Vision<br>• Provide Tutoring and Training<br>• Comprise Employees in the Process<br>• Address Concerns and Objections<br>• Lead by Example<br>• Celebrate Successes and Milestones<br>• Provide Ongoing Support and Resources<br>• Create a Culture of Continuous Improvement<br>• Monitor Progress and Adjustments. (Almeida et al. 2020) |

**TABLE 5.2 (Continued)**
**List of challenges and their respective ways to overcome it**

| S. No | Challenges | Ways to overcome |
|---|---|---|
| 5. | Integration and Interoperability | • Define Clear Integration Goals<br>• Standardize Data Formats and Protocols<br>• Invest in Integration Platforms and Middleware<br>• Implement Data Governance and Master Data Management (MDM)<br>• Use APIs and Microservices Architecture<br>• Conduct Comprehensive Testing and Validation<br>• Collaborate with Stakeholders and Partners<br>• Monitor and Maintain Integration Solutions (Ok et al. 2021) |
| 6. | Cost and Return on Investment (ROI) | • Define Clear Objectives and KPIs<br>• Conduct Cost-Benefit Analysis<br>• Prioritize High-Impact Projects<br>• Phased Approach and Quick Wins<br>• Explore Cost-Effective Solutions<br>• Optimize Resource Allocation<br>• Measure and Monitor Performance<br>• Continuous Improvement and Iteration (Vita et al. 2021) |
| 7. | Regulatory and Compliance Issues | • Stay Informed and Updated<br>• Conduct Regulatory Impact Assessments<br>• Engage Legal and Compliance Experts<br>• Establish Compliance Governance Framework<br>• Implement Compliance-by-Design Principles<br>• Document Policies and Procedures<br>• Conduct Regular Compliance Audits<br>• Provide Training and Awareness Programs<br>• Monitor Regulatory Changes and Updates (Apostolaros et al. 2020) |
| 8. | Digital Divide and Accessibility | • Understand the Digital Divide<br>• Expand Internet Access<br>• Provide Affordable Devices and Services<br>• Promote Digital Literacy<br>• Design Accessible Digital Products<br>• Tailor Content for Local Communities<br>• Partner with Community Organizations<br>• Advocate for Policy Changes<br>• Measure Impact and Progress (Nayyar et al. 2019) |

## 5.5 DIGITALIZATION STRATEGIES FOR SCALING

An integral part of digitalization strategies of high-growth efforts is using the new era of digital technologies and processes to drive growth, efficiency, and innovation. Scale technology infrastructure (i.e., cloud computing, etc.) allows businesses to extend their reach and acquire clients via digital marketing channels. Enabling direct sales with e-commerce platforms and streamlined online sales channels, alongside data-driven decision-making and process automation, enhances operational efficiency. Consumer experience, digital collaboration tools, and agile operations are additional scalable and adaptive functionalities. Strategic partnership, continuous learning, and an innovative culture are key to helping the organizations stay competitive in the digital economy. Utilizing digital technologies and processes to enable growth efficiency, and innovation is essential for scaling. The following are some of the important techniques to consider:

a. **Scalable Technology Infrastructure:** Cloud computing and other scalable technologies infrastructure is essential if one aims to grow business and deploy new technologies without a huge upfront investment in hardware or infrastructure. With the flexibility, scalability, and agility provided by cloud-based solutions, organizations can adapt to meet their evolving needs cost-effectively (Westerlund 2020)
b. **Digital Marketing and Customer Acquisition:** Utilizing email advertising, social broadcasting marketing, SEO, and gratified marketing are the different ways to reach people and attract new customers. Target specific consumer segments through data-driven marketing and personalized messaging for improved engagement (Westerlund 2020)
c. **E-commerce and Online Sales Channels:** The expansion of the sales channels can be performed by creating an online marketplace or e-commerce platform to sell goods and services directly to customers. The probability of increasing sales and revenue is performed by providing a user-friendly shopping experience, trusted payment methods, and personalized recommendations (Buheji, 2018)
d. **Data-driven decision-making:** By utilizing data analytics and tools like business intelligence, one can gain insights into the decision-making information. The utilization of data to find new opportunities and optimize operations will enhance client interactions. One can stay ahead by using machine learning algorithms, and predictive analytics to make proactive decisions based on client preferences (Cantamessa et al. 2020)
e. **Process Automation and Efficiency:** Tools such as robotic process automation and workflow automation help in automating repetitive tasks, helping companies streamline processes. Automating manual processes increases productivity, reduces errors, and allows resources to be used to focus on higher-value tasks. Determining the departments such as finance, HR, and customer service for automating the task will promote productivity and scalability (Priyanto et al. 2023)
f. **Agile and Flexible Operations:** To respond swiftly to shifting consumer needs and market situations, implement agile approaches and flexible operational

procedures. Adopt an innovative, experimental, and continuous improvement culture to promote responsiveness and agility throughout the entire company. Use agile project management methodologies to promote cooperation and iterative development, such as Scrum or Kanban (Mula et al. 2024)

g. **Digital Collaboration and Communication:** It allows a distributed remote workflow to work from anywhere in the world, including things such as project management software, team collaboration platforms, and video conferencing solutions. It promotes collaboration, knowledge sharing, and instant communications in a digital workplace to increase productivity and team play (Gartner et al. 2024)
h. **Customer Experience Enhancement:** To improve the customer, experience several tools can be utilized such as chatbots, smart applications, and self-service portals, etc. In addition, customer relationship management solutions enable to management of relationships, monitor all interactions that take place between customer interaction, and increase loyalty and retention of customers as well (Holmlund et al., 2020)
i. **Partnerships and Ecosystems:** It helps in exploring and forming strategic partnerships along with startups and other organizations that have access to new markets, skills, and resources. Build an ecosystem of suppliers, collaborators, and stakeholders to extend the offerings with complementary solutions (AKSOY, 2023)
j. **Continuous Learning and Innovation:** Foster change, flexibility, and a mindset of continual improvement in the organization are the keys for continuous learning. Encouraging the staff members to adopt the new technologies, acquire digital know-how and embrace change as a growth opportunity. Invest in training and development programs for staff members to create a knowledgeable workforce that can spearhead projects and drive digital transformation (Sjödin et al., 2021).

### 5.5.1 Insights on Scaling Digital Infrastructure, Workforce, and Operations

To enable growth, novelty, and competitiveness in today's digital economy, scaling the digital infrastructure, workforce, and operations calls for a comprehensive strategy that incorporates technological investments, people development, and process optimization. Businesses that put scalability and agility first in their digital activities will be in a better position to take advantage of new opportunities and successfully negotiate changing market dynamics. A comprehensive approach is needed to scale digital infrastructure, personnel, and operations to support innovation, encourage growth, and drive efficiency (represented in Figure 5.6).

The definition of "scale," considering the viewpoints of end users, patients, healthcare policymakers, investors, and unmet needs within the pertinent healthcare systems, is a prerequisite for evaluating the effectiveness of growing digital health efforts. When a digital solution is smoothly integrated into the healthcare system rather than being viewed as a stand-alone endeavour, it could be one definition of effective scale. Being able to scale digital solutions quickly sets firms that gain value from

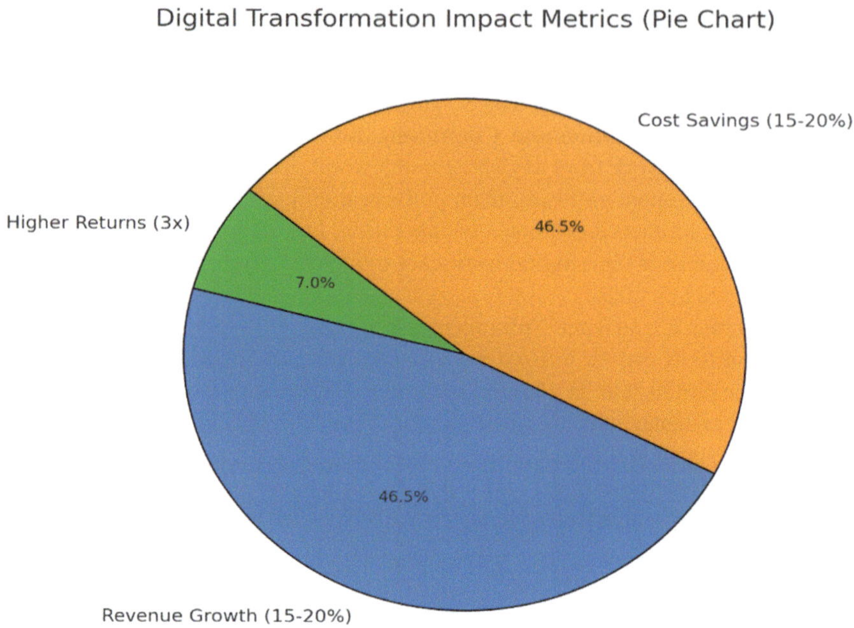


**FIGURE 5.6** Financial benefits of business after opting for Digitalization. (Sabbagh et al. 2012.)

their digital transformations apart from those that do not. Scalable digital solutions are key to creating meaningful value. In periods of economic or company uncertainty, the need to scale successfully increases as shown in figure 5.6. Businesses that use digital solutions at scale enhance long-term resilience while achieving short-term improvements in income, cost, and speed.

## 5.6 FUTURE TRENDS IN DIGITALIZATION

Digitalization will keep transforming industries and changing how businesses operate in the years to come. Developments in machine learning and artificial intelligence (AI), which enable automation and data-driven decision-making, are important themes. The 5G network will enhance the connectivity of services like argument reality and virtual reality, the devices based on the IoT help to populate individuals' lives with more relevant real-time data to work from, facilitating a better process. The concerns related to privacy and cybersecurity will only continue to be evaluated, while edge-computing will enable real-time processing and lower latency. Blockchain technology will change transaction security and transparency, and digital health solutions and telemedicine will modernize healthcare accessibility. Digitalization will further integrate sustainability and green technologies, encouraging environmentally responsible

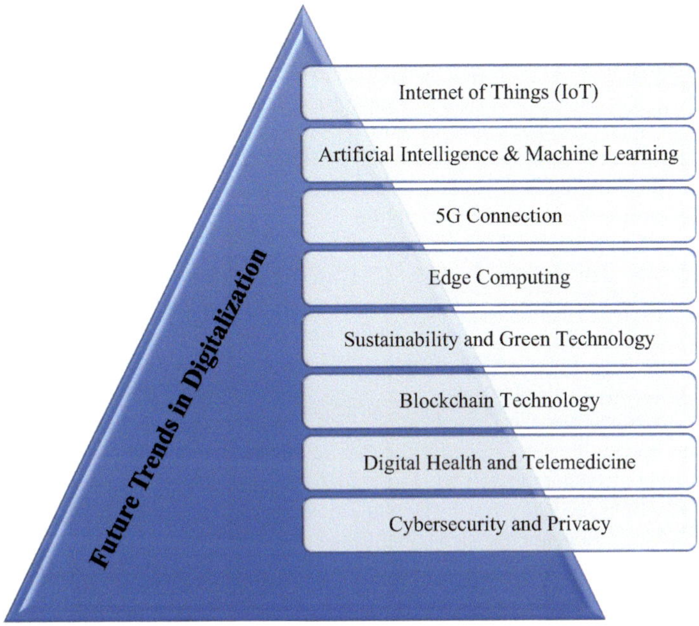


**FIGURE 5.7** Future Trends in Digitalization.

behaviours. Businesses that deliberately adopt these trends will be better equipped to flourish in the quickly changing digital environment. The diagrammatic view of the future trends in digitalization is shown in the Figure 5.7.

## 5.7 CONCLUSION

Businesses will have countless chances to innovate, expand, and adjust to a quickly changing environment as a result of the future of digitalization. Organizations must be alert in utilizing the latest developments in AI, IoT, 5G, and other emerging technologies to boost productivity, improve customer experiences, and keep a competitive edge. Nevertheless, despite the allure of digitalization, companies must place a high priority on sustainability, data privacy, and cybersecurity to guarantee the ethical and responsible use of digital technology. Through strategic adoption of digitization and proactive problem-solving, organizations may set themselves up for success in the dynamic and interconnected world of the future. The path of alphanumeric transformation, looking ahead, will be characterized by ongoing learning, innovation, and adaptation, profoundly and dramatically influencing the future of business.

## REFERENCES

Aksoy, C.: Digital Business Ecosystems: An Environment of Collaboration, Innovation, and Value Creation in the Digital Age. *J. Bus. Trade*. 4, 156–180 (2023). https://doi.org/10.58767/joinbat.1358560

Ali, T., Al-Khalidi, M., Al-Zaidi, R.: Information Security Risk Assessment Methods in Cloud Computing: Comprehensive Review. *J. Comput. Inf. Syst.* 00, 1–28 (2024). https://doi.org/10.1080/08874417.2024.2329985

Almeida, F., Duarte Santos, J., Augusto Monteiro, J.: The Challenges and Opportunities in the Digitalization of Companies in a Post-COVID-19 World. *IEEE Eng. Manag. Rev.* 48, 97–103 (2020). https://doi.org/10.1109/EMR.2020.3013206

Anthony Jnr, B.: Enabling Interoperable Distributed Ledger Technology with Legacy Platforms for Enterprise Digitalization. *Enterp. Inf. Syst.* 18, 1–33 (2024). https://doi.org/10.1080/17517575.2023.2255979

Apostolaros, M., Babaian, D., Corneli, A., Forrest, A., Hamre, G., Hewett, J., Podolsky, L., Popat, V., Randall, P.: Legal, Regulatory, and Practical Issues to Consider When Adopting Decentralized Clinical Trials: Recommendations From the Clinical Trials Transformation Initiative. *Ther. Innov. Regul. Sci.* 54, 779–787 (2020). https://doi.org/10.1007/s43441-019-00006-4

Berger, T., Benedikt Frey, C.: Digitalization, jobs, and convergence in Europe: strategies for closing the skills gap. Prepared for the European Commission DG Internal Market, Industry. *Oxford Martin Sch.* 51, 1–51 (2016).

Bröckling, G., Behr, J., Erdmann, J.: Digital Transformation of an Educational Organization from a Media Educational Viewpoint. Cham, 1–23 (2021).

Buheji, M.: To Cite This Version: HAL Id: halshs-01820594. *Issues Soc. Sci.* 6, 1–8 (2018).

Cantamessa, M., Montagna, F., Altavilla, S., Casagrande-Seretti, A.: Data-Driven Design: The New Challenges of Digitalization on Product Design and Development. *Des. Sci.* 6, 1–33 (2020).https://doi.org/10.1017/dsj.2020.25

Colding, J., Nilsson, C., Sjöberg, S.: Smart Cities for All? Bridging Digital Divides for Socially Sustainable and Inclusive Cities. *Smart Cities.* 7, 1044–1059 (2024). https://doi.org/10.3390/smartcities7030044

Gartner, J., Maresch, D., Tierney, R.: The Key to Scaling in the Digital Era: Simultaneous Automation, Individualization and Interdisciplinarity. *J. Small Bus. Manag.* 62, 628–655 (2024). https://doi.org/10.1080/00472778.2022.2073361

Hidayat, M., Defitri, S.Y.: Digitalization and the Changing Landscape of Tax Compliance (Challenges and Opportunities). *Account. Stud. Tax J.* 1, 131–139 (2024). https://doi.org/10.62207/c2gyc030

Holmlund, M., Van Vaerenbergh, Y., Ciuchita, R., Ravald, A., Sarantopoulos, P., Ordenes, F.V., Zaki, M.: Customer Experience Management in the Age of Big Data Analytics: A Strategic Framework. *J. Bus. Res.* 116, 356–365 (2020). https://doi.org/10.1016/j.jbusres.2020.01.022

Kraus, S., Jones, P., Kailer, N., Weinmann, A., Chaparro-Banegas, N., Roig-Tierno, N.: Digital Transformation: An Overview of the Current State of the Art of Research. *SAGE Open.* 11, 1–15 (2021). https://doi.org/10.1177/21582440211047576

Lichtenthaler, U.: The role of top inventive organizations in the digital age. *Res. Technol. Manag.* 61(4), 27–35 (2018).

Mardiani, E., Utami, E.Y., Farooq Mujahid, M.U.: B2B Digital Marketing and ROI Measurement: Challenges and Opportunities in the Business-to-Business Industry for MSMEs in Indonesia. *West Sci. Interdiscip. Stud.* 1, 867–875 (2023). https://doi.org/10.58812/wsis.v1i09.249

Mula, C., Zybura, N., Hipp, T.: From Digitalized Start-up to Scale-up: Opening the Black Box of Scaling in Digitalized Firms towards a Scaling Process Framework. *Technol. Forecast. Soc. Change.* 202, 123275 (2024). https://doi.org/10.1016/j.techfore.2024.123275

Nayyar, T., Aggarwal, S.S.S., Khatter, D., Kumar, K., Goswami, S., Saini, L.: Opportunities and Challenges in Digital Literacy: Assessing the Impact of Digital Literacy Training for Empowering Urban Poor Women. *University of Delhi*. 160, 33–42 (2019).

Ok, J.S., Kwon, S.D., Heo, C.E., Suh, Y.K.: A Survey of Industrial Internet of Things Platforms for Establishing Centralized Data-Acquisition Middleware: Categorization, Experiment, and Challenges. *Sci. Program.* 2021, 5598025, 1–13 (2021). https://doi.org/10.1155/2021/6641562

Pan, S., Trentesaux, D., McFarlane, D., Montreuil, B., Ballot, E., Huang, G.Q.: Digital Interoperability in Logistics and Supply Chain Management: State-of-the-Art and Research Avenues towards Physical Internet. *Comput. Ind.* 128, 1–20 (2021). https://doi.org/10.1016/j.compind.2021.103435

Pau, M., Mirz, M., Dinkelbach, J., McKeever, P., Ponci, F., Monti, A.: A Service Oriented Architecture for the Digitalization and Automation of Distribution Grids. *IEEE Access*. 10, 37050–37063 (2022). https://doi.org/10.1109/ACCESS.2022.3164393

Priyanto, P., Murwaningsari, E., Augustine, Y.: Exploring the Relationship between Robotic Process Automation, Digital Business Strategy and Competitive Advantage in Banking Industry. *J. Syst. Manag. Sci.* 13, 290–305 (2023). https://doi.org/10.33168/JSMS.2023.0320

Regulations, A.M.L.C.F.T., Suleh-yusuf, M.: Big Data, Internet Privacy and the Vulnerabilities of the African Regulatory Landscape. *Eur. J. Bus. Manag.* 12(17), 127–141 (2020). https://doi.org/10.7176/ejbm/12-17-14

Sabbagh, K., Friedrich, R., El-Darwiche, B., Singh, M., Ganediwalla, S.: Maximizing the Impact of Digitization. *Glob. Inf. Technol. Rep.* 2012. 121–133 (2012).

Sjödin, D., Parida, V., Palmié, M., Wincent, J.: How AI Capabilities Enable Business Model Innovation: Scaling AI through Co-evolutionary Processes and Feedback Loops. *J. Bus. Res.* 134, 574–587 (2021). https://doi.org/10.1016/j.jbusres.2021.05.009

Sundermeier, J., Birkner, S., Ettl, K., Kensbock, J.M., Tegtmeier, S.: Hello Diversity! Opportunities and Challenges of Entrepreneurial Diversity in the Digital Age. *Commun. Assoc. Inf. Syst.* 47, 697–715 (2020). https://doi.org/10.17705/1CAIS.04732

Vasilev, V.L., Gapsalamov, A.R., Akhmetshin, E.M., Bochkareva, T.N., Yumashev, A.V., Anisimova, T.I.: Digitalization Peculiarities of Organizations: A Case Study. *Entrep. Sustain. Issues.* 7, 3173–3190 (2020). https://doi.org/10.9770/jesi.2020.7.4(39)

Vita, V., Christodoulou, C., Zafeiropoulos, I., Gonos, I., Asprou, M., Kyriakides, E.: Evaluating the Flexibility Benefits of Smart Grid Innovations in Transmission Networks. *Appl. Sci.* 11, 1–18 (2021). https://doi.org/10.3390/app112210692

Westerlund, M.: Digitalization, Internationalization and Scaling of Online SMEs. *Technol. Innov. Manag. Rev.* 10, 48–57 (2020). https://doi.org/10.22215/TIMREVIEW/1346

# 6 Navigating Challenges for Tech-Based Enterprises Using Multi-Criteria Decision-Making Approach

*Rajender Kumar, Sushant, Rahul Sindhwani, Punj Lata Singh, Harish Kumar Banga, and Raman Kumar*

## 6.1 INTRODUCTION

Going on an entrepreneurial journey is a present-day need and an essential topic for debate for academicians and practitioners. Almost all developing economies emphasize promoting entrepreneurial activities. From the nationwide growth perspective, creating wealth is the most challenging task and becomes more complex as it is confined to a few sectors. Traditionally, agriculture and manufacturing (including assembling and trading) sectors were the key drivers for economic growth. But now, things are changing, and e-businesses and the contribution of startups' is also recognized. These startups are represented as newly carved entities in a highly dynamic and uncertain environment with rapidly evolving markets and technologies. Besides contributing to economic growth, these kinds of startups are significantly contributing and recognized for their role in generating employment (Ardianti and Inggrid, 2018).

The primary goal for initiating the entrepreneurial journey is to build a sustainable business model that provides the necessary support for nationwide growth. The present-day market has more dynamism than the earlier years and has led to a competitive environment that could catalyze startups' emergence at national and global levels. The climate of innovation, quality, and customer centrism has provided the ideal breeding ground for the emergence of startups Although researchers have proposed different definitions of the term over time, the most widely used definition is one given: "Startups are a human institution designed to create new products and services under conditions of extreme uncertainty. The startups focus on creating demand by introducing new business models or solving emerging problems of society. The

 DOI: 10.1201/9781003537335-6

world economic forum has differentiated between SMEs and startups as the former do not indulge in innovation to run and scale the enterprise" (Yang, 2019).

To distinguish it from small businesses, startups keep a business model or innovative technology at their core and focus on quick growth. The startups that grow quickly are easily scalable and can be expanded with low incremental cost using technology and innovation. The literature on entrepreneurship reveals that exponential growth has been reported in amount of new innovative businesses in the past decade. The rapid advancements in this area have led to an increased number of IT startups addressing any uncertainty through better forecast mechanisms. This development has been accompanied by the growth of low-cost operations and the presence of a support system for startups known as incubators. The contemporary economic landscape has motivated young startups to concentrate on disruptive technologies to maintain competitiveness (Supardianto, 2019).

The IT infrastructure revolutionized the information processing system, which helped many startups grow rapidly. The IT-enabled channels helped startups in designing and developing products; creating niches; managing the supply according to demand; estimating product pricing based on uncertainties; promoting the products, etc. While reviewing this from a societal perspective, the technology-based startups directly compete in the market individually. Further to this, the technology helps to make the idea viable by integrating customer perspectives and potential investors. The evolution of technology such as blockchain, machine learning, and artificial intelligence has disrupted the existing ecosystem and has supported IT-enabled startups to compete worldwide (Sunny, 2019).

The common agenda for associating the IT-enabled channels with energizing the startup activities is shown in Figure 6.1.

Despite contributing toward nation growth, here, it is an exaggeration to state that only 10–20% of startups have the potential for high growth and scalability In India, over 5000 technology-based startups are currently in operation with a success rate

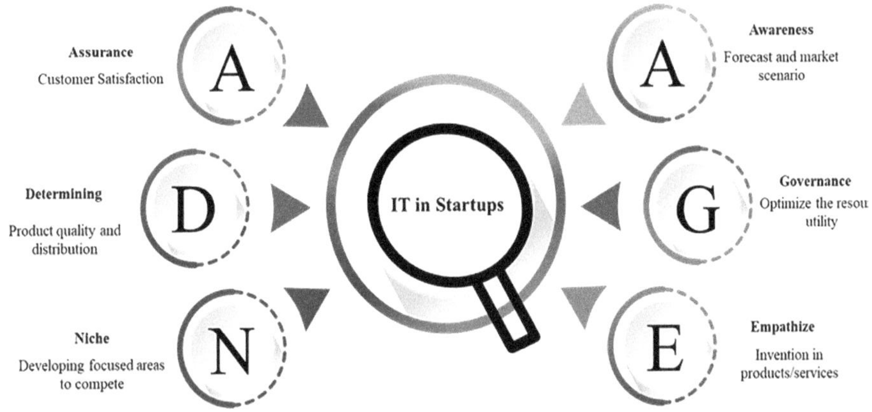


**FIGURE 6.1** Agenda for associating IT-enabled channels in startups.

of 10–15% only. The failure rate of startups is very high worldwide. Against this backdrop, it is very pertinent to understand and to answer the question what are the possible causes for most startups failures, and how this can be avoided or eliminated. The recent literature on startups reveals that the inconsistency in the functioning of various operations of startups is especially related to product selection and development (Santisteban et al., 2021).

Despite the immense potential of such startups, only a few managed to grow at a decent, sustainable pace. It was also observed that the entrepreneurs who rely on the IT-enabled tools sometimes ignore important business activities as they do not see this as part of their job. Based on this proposition, the present work seeks to explore the role and contribution of IT-enabled systems in performing various entrepreneurial functions. The present work is aimed to provide a better understanding of the contribution of IT-enabled systems in managing and operating startups effectively (Rafiki, 2024).

This paper contributes to the literature by achieving the following objectives:

- To explore the necessary activities of startups that can be managed effectively and efficiently through IT-enabled channels
- To examine the cause-and-effect relations among the identified activities
- To rank all identified activities based on their relative importance.

The present work includes defining sixteen essential activities of the startups that directly contribute to accelerating the growth. To define such activities, an extensive literature review is carried out. In addition, the relationships among identified activities are analyzed to define each activity's priority using MCDM approaches.

## 6.2 LITERATURE REVIEW

The statistics on startup failures worldwide reveal that around 20% failed in the first year, around 30% in the next year, and around 50% within five years of inception. The entrepreneurs, who fail to run startups, never go through the facts as they rely on the idea (problem to solve), which is the toughest part because no one can go for a startup with only an idea. The failed startups commonly consider the lack of financial management as the primary cause. Though managing finance in startups is crucial and directly impacts the functioning of operations, the share of financial mismanagement in the failure of startups is not more than 20%. Other factors are not highlighted in earlier studies, such as people's selection (team members/partners), market awareness, customer perception, regulatory compliances, and building brands. Meeting customer expectations is also a concern for startups. The literature also revealed that startups are promoting their products and services. Often, they have failed to highlight the pipelined activities for the future. Here, it is an exaggeration to say that the involvement of technology contributes a lot to managing and upgrading the startup's ability to communicate with customers directly and address the critical concerns related to products and deliveries in terms of cost, timing, feature, and coverage greater parts of the market (Satyanarayana, 2021).

The earlier studies on the startups explored and examined the various activities critically in the startups. The main emphasis of most of the reports was on the financial activities and the process of delivering services/products. The research gap perceived for the present study explores all essential activities for future startups and how the essential activities can contribute to startups functioning. Existing studies have also confirmed using IT-based tools for fewer activities, or for product design and development, delivery and customer feedback/demand, etc. Proper planning and execution are required to manage all the activities of startups simultaneously.

### 6.2.1 Identification of Essential Activities in Entrepreneurial Setups

A rapid increase in entrepreneurial activities would further contribute to wealth creation through channelized operations. The Indian economy has also witnessed a sharp rise in startups over the past few years and this has started playing an important role in the country's economic growth. The main focus of startups is innovation (not tested before) and the growth perspective (exponential instead of linear). For this, they rely more on technology in managing the operations such as demand forecasting, procurement, sourcing, manufacturing, distribution, etc. Involving technology, e.g. IT-enabled tools, in the smooth functioning of the startup operations prepared the growth perspective's footprints. Undoubtedly, almost all technology-based startups rely on IT-enabled systems, which creates value for them from purchasing stage to selling (Rippa, 2019). This study identified the operations in the present-day business models that allow startups to grow rapidly. The identified activities are listed in Table 6.1.

**TABLE 6.1**
**List of identified essential activities for startups**

| Sr. No. | Activity | Objectives | Justification |
|---|---|---|---|
| 1 | Problem (Idea/ Opportunity) Identification | To justify the problem found with the possible alternatives that already exist | The problem is also referred to as opportunity. In start-ups, IT-enabled channels help get concise information on future predictions through market trends, feedback, competitors' strategies, etc. |
| 2 | Defining the Purpose | To define the objective of the entrepreneurial setups to specify the market segment | The IT-enabled channels help define the startups' clear-cut purposes and create the brand image among customers. |
| 3 | Planning for Entrepreneurial Operations | To design and develop the plans for different operations regarding the resource's optimization | It is not compulsory that every startup has all the resources and is utilized satisfactorily. The IT-enabled channels help in systematic planning and further strategizing the startup's resource utility. |

*(continued)*

**TABLE 6.1 (Continued)**
**List of identified essential activities for startups**

| Sr. No. | Activity | Objectives | Justification |
|---|---|---|---|
| 4 | Product Design and Development | To discuss the product characteristics that enable the customer to buy it for the sake of value | The IT-enabled channels provide advanced solutions to engineering designs because it is a must for every startup to validate a product before the launch. |
| 5 | Positioning of finished products/ services | To specify the perfect information regarding the deliberation of product features among customers | It is one of the critical factors which help young startups to reach the intended audience. The IT-enabled channels help in positioning the end products in the customer's mind by providing the information needed to make the final purchase. |
| 6 | Proposition of value | To define the uniqueness (Point of difference) which motivates a customer to buy your product in comparison with others | The startups should not be carried away by buzzwords or meaningless slogans but should deliver a unique value proposition to their audience. IT-enabled channels help create and leverage the values among the products/services offered. |
| 7 | Prospecting the future | To find the opportunities that best match with company's core competencies | Another crucial step is prospecting, which involves finding new prospects through advertising, cold calling, public relations, and other marketing tactics for new products and services.<br>Here, IT-enabled channels such as social media platforms help startups to stay connected with their existing customers and get real-time feedback. |
| 8 | Price Determination | To calculate the ROI through the product price determination | Pricing helps in the acceptability of the product to the customers and profitability of the product to the young IT startups. It is the most flexible instrument that can influence these startups' sales volume and market share. IT-enabled channels provide agility to the system to counter competition in the market by adjusting the price to meet the customer requisition based on real-time data. |

**TABLE 6.1 (Continued)**
**List of identified essential activities for startups**

| Sr. No. | Activity | Objectives | Justification |
|---|---|---|---|
| 9 | Packaging attractiveness | To suggest the product packaging that is best suited to the brand by comparing and trending | Since packaging has the potential to attract new customers and retain existing ones, aesthetic aspects of packaging can act as a marketing tool for startups. The IT-enabled channels contribute to designing the packaging of the products. |
| 10 | Place `selection for operating the setups | To determine the place from where the industry will operate | Conveniently located startups are better placed to attract more customers and improve their visibility. The place is important as it can facilitate fast business operations, and suppliers can expediently reach to deliver the raw materials on time. It also helps make the product available in the market at a lesser cost and time frame. |
| 11 | Promotion Activities | Describe how to set out the brand value in the customer's mind. | Young startups need multiple promotional activities like personal selling, sales promotion, advertising, and public relations to entice customers, generate sales and boost customer engagement. |
| 12 | Engaging Partnership | To find out the partners and the centre of influence which will help in creating the referrals for the start-ups | The innovation and entrepreneurship ecosystem are the relationship between public and private entities and entrepreneurs, facilitating the success of emerging technology-based companies. This ecosystem offers many advantages to innovative companies, including access to funding sources, trained personnel, and business networks. |
| 13 | Pipelining the Activities | To qualify the industry prospects, keep them warm and build the relationship until the start-up finds the targeted audience becomes the client | Pipeline employs strategies like inbound marketing, growth hacking, and lead nurturing to connect sales and marketing to information. IT-enabled channels help develop the pipeline's list of activities (amendments to existing products; new products, new market segments, etc.) |

*(continued)*

**TABLE 6.1 (Continued)**
**List of identified essential activities for startups**

| Sr. No. | Activity | Objectives | Justification |
|---|---|---|---|
| 14 | People Management | To find out the members who dedicatedly devote their talent and skill to take the business on the way to success | A set of resources are required to drive the entrepreneurial venture. The IT startups require individual abilities, skills, and access to information and resources. They help cultivate a conducive environment, capital, networks, and mentorship. The right human resources is also a precondition to succeed for these young startups. |
| 15 | Performance Analysis and Evaluation | To define the specific measures which help in controlling the industrial operations and decisions as well | In the present scenario, higher performance is the main requisite in delivering a high-value product and/ or service to the customer.<br>For startups, performing consistently to overcome changing customer needs is very important. The IT-enabled channels support the tools used for monitoring and evaluating performance. |
| 16 | Persistence to Reisolate | To prepare the plan to stay focused on the target by accurately estimating future events. | The IT enables channels to provide estimates on the performance and even future events. The inputs motivate the startups to reisolate the Improved performance even during disruption. |

## 6.3 RESEARCH METHODOLOGY

The present study has been conducted in four phases, as shown in Figure.6.2. The first phase begins with the exploratory literature review through which the sixteen essential activities for startups are identified as listed in Table 6.1 earlier. In the second phase, the data collection and validation through experts are done. The third phase of the study is a very crucial one. In this phase, the MCDM approaches such as Entropy, MOORA, and VIKOR are applied to data collected to accomplish research objectives. At last, the outcomes from the various approaches are discussed and summarized systematically.

The present work aims to examine the contribution of essential activities in the growth of startups when these are managed through IT tools. Four major criteria are subdivided into two sections for exploring the contribution, i.e., beneficial and non-beneficial, and explained as follows:

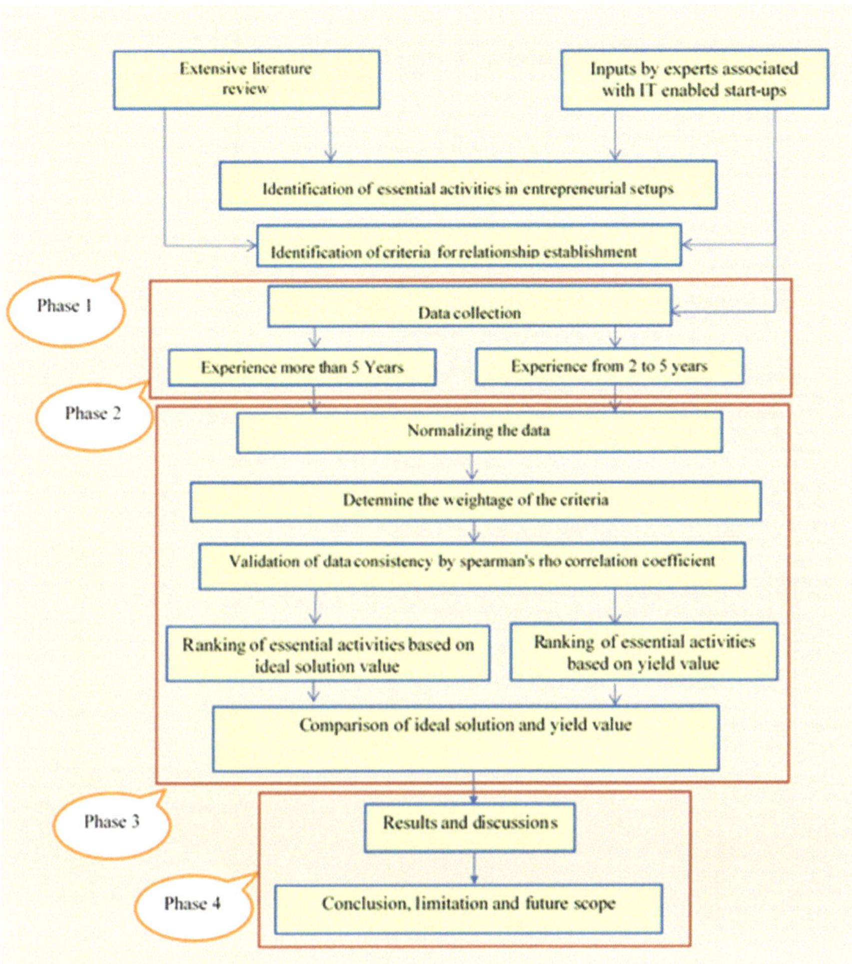


**FIGURE 6.2** Four phased research frameworks.

- **Product improvement (C1):** The essence of all IT-based startups is the product improvement. The real challenge faced by the entrepreneur is to break the barriers of inertia if the product is already established in the market. Further, dynamic capabilities are required so that the startups can easily integrate/construct internal and external competencies to quickly address the changing market conditions and provide fruitful solutions
- **Customer satisfaction (C2):** It is a massive task to address customer satisfaction with scarce resources at their disposal. There is a need to use the resources optimally, which means the product should be designed according to the customer's requirements. Here, quality is greatly valued and allows a company to become successful. The primary goal of all companies, and the

key to sustainable growth, is that their customers will get satisfaction from the services/products offered

- **Financial obligation (C3):** A set of financial instruments provided by different actors, including local government, angel investors, and venture capitalists, is required to run the startups. Funding organizations with innovative business models is fundamental for businesses' long-term survival and growth
- **Resource requirement (C4):** Various resources are required, like business incubators. Business incubators facilitate the formation of successful startups. They help launch innovative products/ services in the market and also help in getting access to seed money, advice, and mentoring support during the initial years.

### 6.3.1 Determine the Weightage of the Criteria

In this study, the Entropy approach is used to determine the weightage of the criteria. It helps define the system's characteristics, whether in order or a lack of order, based on the attributes. It is like the thermodynamic system, which states that it is the way to measure the presence or absence of order in a system. This was first used by to explore the information systems performance based on criteria selected for the study. It is widely used for finding solutions to real-life problems where more than one criterion is used, such as supplier selection and product evaluation. This approach more relevant in the studies where the decisions are to be made based on measuring relative importance among each criterion under study. Compared with other approaches available under the umbrella of MCDM, this approach is simpler to understand and applied easily through mathematical calculations. Even the time took to get the outcomes are less in this approach. The Fuzzy-Delphi approach is used to overcome the problem of deriving the outcome based on experts having varied expertise (Leinster, 2021).

The steps in the Entropy approach are as follows:

**Step 1: Data collection:** All four criteria discussed above are used to examine the system characteristics to detect inconsistency. Based on this criterion, the inputs are collected from the experts. Here, the targeted entrepreneurs are divided into two groups, i.e., the first group comprises the entrepreneurs with functional startups of more than five years; and the second group where the entrepreneurs have been in operation for between two to five years, are interviewed. A total of 12 experts agreed to participate in the study based on convenience sampling. The entrepreneurs interviewed here belong to the engineering community specializing in computer science, information technology, mechanical engineering, and electronics, and communications.

**Step 2: Normalizing of a matrix:** In the qualitative analysis, the normalizing of the inputs is done on an individual basis. It helps eliminate any ambiguity in the data pattern. In this study, the data is normalized using Equation (6.1).

$$X_{ij} = \frac{Xij}{\sum_{i=1}^{m} Xij} (\mathrm{j} = 1,2,3 \ldots \mathrm{n}) \tag{6.1}$$

$$N_j = -k\sum_{i=1}^{n} X_{ij} * \ln\left(X_{ij}\right) \qquad j = 1,2.......m \tag{6.2}$$

Where

$$k = 1/\ln\left(n\right) \tag{6.3}$$

**Step 3: Calculation of weight for each criterion:** After normalizing the inputs, the weight for individual criteria '*j*' is calculated by using Equation (6.4).

$$w_j = \frac{1 - N_j}{\sum_{j=1}^{n}\left(1 - N_j\right)} \text{ where } j = 1,\ldots,n. \tag{6.4}$$

### 6.3.2 Examining the Essential Activities Based on Ideal Solution Value

For examining purposes, the VIKOR approach is used on the identified activities. This approach is mainly used to rank alternatives and determine the compromised solutions closest to the ideal. This approach is best suited to explore the solutions to decision-related issues where conflict of interest between the two parties plays an important role. In these types of problems, the decision-maker looks for the alternative closest to the ideal. Therefore, it is assumed that the compromise solution is acceptable to resolute the conflicts. This approach aims to support the maximum utility for the majority of the user and, on the other hand, reduce the opponents' regret level. The approach gained recognition in 2004. The steps for VIKOR analysis are as follows:

**Step 1: Data collection and normalization:** For VIKOR, the input data is required to go ahead. As in the case of ENTROPY, the data provided by the experts is collected and normalized accordingly.

**Step 2: Finding the ideal solution:** An ideal solution means the solution that one can assume, and it may or may not be equivalent to the real solutions as it is based on the number of attributes. In the VIKOR, both the positive and negative ideal solutions are calculated. The positive ideal solution ($E_i$ values) is calculated by using equation (6.5) for beneficial criteria and equation (6.6) for non-beneficial criteria

$$E_i = \sum_{j=1}^{n} W_j \frac{X_{ijmax} - X_{ij}}{X_{ijmax} - X_{ijmin}} \tag{6.5}$$

$$E_i = \sum_{j=1}^{n} W_j \frac{X_{ij} - X_{ijmin}}{X_{ijmax} - X_{ijmin}} \tag{6.6}$$

The negative ideal solution ($F_i$ values) is calculated by using equation (6.7) for beneficial criteria and equation (6.8) for non-beneficial criteria

$$F_i = Maximum\,of \frac{W_j\left(X_{ijmax} - X_{ij}\right)}{X_{ijmax} - X_{ijmin}} (\mathrm{i} = 1, 2, \ldots\ldots, \mathrm{n}) \tag{6.7}$$

$$F_i = Maximum\,of \frac{W_j\left(X_{ij} - X_{ijmin}\right)}{X_{ijmax} - X_{ijmin}} (\mathrm{i} = 1, 2\ldots\ldots, \mathrm{n}) \tag{6.8}$$

**Step 3: Ranking the factors:** Finally, the identified factors are ranked based on the values of $P_i$. The $P_i$ value in the VIKOR analysis is calculated for differentiating among the activities based on effectiveness, and equation (6.9) is used to find the value.

$$P_i = v\left[\frac{E_i - E_{imin}}{E_{imax} - E_{imin}}\right] + (1 - v)\left[\frac{F_i - F_{imin}}{F_{imax} - F_{imin}}\right] \tag{6.9}$$

Here it is noteworthy that the decision-maker assumes the value of the VIKOR constant (v). Finding the outcomes closest to the ideal ones should be kept as 0.5 (Jain, 2020).

### 6.3.3 Examining the Essential Activities based on their Yield Value

After examining the activities through the VIKOR approach, another MCDM named MOORA is applied. The MOORA approach is adopted in solving the problems having the decisions under conflict situations with multiple alternatives. This method was firstly introduced in 2004, and now it is a well-known approach for finding solutions to real-life problems such as material selection, vendor selection, and process optimization. The steps followed in the MOORA method are as follows:

**Step 1: Data collection and normalization:** Similar data is used for the MOORA approach.

**Step 2: Calculating the Yield Value:** The enterprise, in general, has conflicting aims, i.e., it wants to maximize the profit by selling more and more products; on the other hand, it wants to minimize resource consumption. The normalized assessment value is calculated through Equation (6.10) to solve such conflict issues and reported in Table 6.4. It is the value that is added in the case of maximizing and subtracted in the minimization case. In this study, we aim to maximize the weights of beneficial criteria ($C_1$ and $C_2$) and minimize the weights of non-beneficial criteria ($C_3$ and $C_4$). It will also help assess the importance of the factors (activities in the present study) over the others.

$$Y_i = \sum_{j=1}^{g} X_{ij} - \sum_{j=g+1}^{n} X_{ij} \tag{6.10}$$

In Equation 10, the number of beneficial goals is notated by '*g*', and for non-beneficial, it is '*n-g*'.

Further, the importance of each criterion is increased by the multiplication of respective weight to the criteria and calculated through Equation (6.11).

$$Y_i = \sum_{j=1}^{g} Wj*Xij - \sum_{j=g+1}^{n} Wj*Xij \ (j = 1, 2, 3.....n) \tag{6.11}$$

Where, $W_j$ is the weight assigned to $j^{th}$ activity.

**Step 3: Ranking determination:** The ranking in the MOORA approach is then determined with the help of calculated normalized assessment values. Here, it is noteworthy that all values of normalized assessment ($Y_i$) are acceptable. The $Y_i$ value can be positive, negative, or zero as it depends on different criteria (beneficial and non-beneficial in this study) and the total weight, respectively. The difference between the values helps in determining the rank.

### 6.3.4 Comparative Analysis

This study uses a novel approach for finding and validating the study's outcomes. For this purpose, the ranking for various activities associated with the startups is done by VIKOR, and MOORA approaches, respectively. The outcomes of the two approaches are now compared to reveal the input data's consistency and reliability. For the comparison purpose, the values for $Y_i$ (yield value) and $P_i$ (positive ideal solution value) is considered. The common ranking is then allotted to all the activities individually based on the gap between $Y_i$ and $P_i$ values. In general, the factor/enabler which is having a higher value of $Y_i$ and the lower value of $P_i$ are the most suitable/affecting parameter and vice versa.

## 6.4 ANALYSIS

The present section will discuss the analysis based on the application of these approaches.

### 6.4.1 Criteria Weightage Calculation

In this study, the relative importance of the various identified activities is determined using the Entropy approach. The inputs related to the activities are collected from the experts. The two subsets of input data are reported in Table 6.2.

**TABLE 6.2**
**Data collection**

| Identified activities | Entrepreneurs with have the startup in operation for greater than five years | | | | The entrepreneur has the startup in operation between two to five years | | | |
|---|---|---|---|---|---|---|---|---|
| | Beneficial | | Non-beneficial | | Beneficial | | Non-beneficial | |
| | $C_1$ | $C_2$ | $C_3$ | $C_4$ | $C_1$ | $C_2$ | $C_3$ | $C_4$ |
| $P_1$ | 7 | 9 | 6 | 5 | 8 | 9 | 6 | 5 |
| $P_2$ | 3 | 8 | 3 | 3 | 5 | 8 | 3 | 2 |
| $P_3$ | 8 | 7 | 8 | 7 | 9 | 7 | 7 | 6 |
| $P_4$ | 3 | 8 | 7 | 4 | 5 | 8 | 7 | 3 |
| $P_5$ | 2 | 3 | 3 | 3 | 4 | 4 | 2 | 3 |
| $P_6$ | 8 | 7 | 5 | 4 | 9 | 7 | 4 | 4 |
| $P_7$ | 8 | 7 | 4 | 7 | 8 | 7 | 4 | 6 |
| $P_8$ | 3 | 5 | 7 | 7 | 4 | 5 | 6 | 6 |
| $P_9$ | 2 | 6 | 4 | 5 | 3 | 7 | 4 | 4 |
| $P_{10}$ | 5 | 3 | 6 | 6 | 6 | 3 | 6 | 6 |
| $P_{11}$ | 2 | 4 | 6 | 6 | 3 | 4 | 5 | 6 |
| $P_{12}$ | 4 | 3 | 6 | 4 | 5 | 3 | 5 | 2 |
| $P_{13}$ | 2 | 7 | 3 | 5 | 3 | 7 | 3 | 5 |
| $P_{14}$ | 6 | 3 | 6 | 3 | 7 | 4 | 5 | 3 |
| $P_{15}$ | 8 | 4 | 4 | 3 | 7 | 4 | 3 | 3 |
| $P_{16}$ | 3 | 5 | 2 | 4 | 3 | 5 | 1 | 5 |

*C1-Productivity improvement; C2- Customer satisfaction; C3- Financial obligation; C4- Resource requirement*

Data collected is then processed for normalization purposes, i.e., estimated individual activity weight using equation (6.1). In addition, the criteria weight is normalized using equations (6.2) and (6.3). Finally, the weight for each criterion is estimated using equation (6.4) and reported in Table 6.3.

Table 6.3 summarizes the individual activities' weight based on expert opinions and the criteria weights based on standard equations reported in section three of the work. For validation of data consistency, spearman's rho correlation coefficient (*R*) is used. The data sets are estimated with a 95% level of significance (*0.05*) value by using equation (6.12)

$$\rho = 1 - \frac{6\sum_{i=0}^{i=n} d_i^2}{n.\left(n^2 - 1\right)} \tag{6.12}$$

In equation (6.12), the *i* and *n* are the number of alternatives. The $d_i$ in the equation represents the difference between the ranks obtained by the two data sets. It is safe to accept that the two data sets are statistically similar, and any of the two can be used for further analyses. The calculated ρ value for the present work is 0.9, which shows the strong relationship and acceptance of the collected data. For further calculation, the second group (entrepreneurs having the startup in operation between two to five years) is considered. Also, the group of second entrepreneurs has a flexible approach and a space for innovation compared to matured entrepreneurs (having expertise of more than five years). The input data is then normalized to remove all the inconsistencies in the data sets.

### 6.4.2 Ranking of Essential Activities Based on Ideal Solution Value

The weights to individual activity and criterion estimated through the Entropy approach is further used for ranking purpose for all individual activities of the startups. The VIKOR approach is used to estimate the positive ideal solution and negative ideal solution for both the beneficial as well as non-beneficial criteria. In Table 6.3, the data estimated for the positive ideal solution (using equations (6.5) and (6.6)) and the negative ideal solution (using equations (6.7) and (6.8)) is reported. In addition, the effectiveness of all individual activities $P_i$ value is estimated using equation (6.9). Based on the $P_i$ values, the ranking is assigned to all activities, as shown in Table 6.4.

**TABLE 6.3**
**Normalized data**

| Identified activities | Entrepreneur's with having the startup in operation for greater than five years | | | | Entrepreneur has the startup in operation between two to five years | | | |
|---|---|---|---|---|---|---|---|---|
| | Beneficial | | Non-beneficial | | Beneficial | | Non-beneficial | |
| | C1 | C2 | C3 | C4 | C1 | C2 | C3 | C4 |
| $P_1$ | 0.18 | 0.18 | 0.17 | 0.15 | 0.17 | 0.18 | 0.18 | 0.17 |
| $P_2$ | 0.08 | 0.16 | 0.08 | 0.09 | 0.10 | 0.16 | 0.09 | 0.07 |
| $P_3$ | 0.21 | 0.14 | 0.22 | 0.21 | 0.19 | 0.14 | 0.21 | 0.21 |
| $P_4$ | 0.08 | 0.16 | 0.19 | 0.12 | 0.10 | 0.16 | 0.21 | 0.10 |
| $P_5$ | 0.05 | 0.06 | 0.08 | 0.09 | 0.08 | 0.08 | 0.06 | 0.10 |
| $P_6$ | 0.21 | 0.14 | 0.14 | 0.12 | 0.19 | 0.14 | 0.12 | 0.14 |
| $P_7$ | 0.21 | 0.14 | 0.11 | 0.21 | 0.17 | 0.14 | 0.12 | 0.21 |
| $P_8$ | 0.08 | 0.10 | 0.19 | 0.21 | 0.08 | 0.10 | 0.18 | 0.21 |
| $P_9$ | 0.05 | 0.12 | 0.11 | 0.15 | 0.06 | 0.14 | 0.12 | 0.14 |
| $P_{10}$ | 0.13 | 0.06 | 0.17 | 0.18 | 0.13 | 0.06 | 0.18 | 0.21 |
| $P_{11}$ | 0.05 | 0.08 | 0.17 | 0.18 | 0.06 | 0.08 | 0.15 | 0.21 |
| $P_{12}$ | 0.10 | 0.06 | 0.17 | 0.12 | 0.10 | 0.06 | 0.15 | 0.07 |
| $P_{13}$ | 0.05 | 0.14 | 0.08 | 0.15 | 0.06 | 0.14 | 0.09 | 0.17 |
| $P_{14}$ | 0.15 | 0.06 | 0.17 | 0.09 | 0.15 | 0.08 | 0.15 | 0.10 |
| $P_{15}$ | 0.21 | 0.08 | 0.11 | 0.09 | 0.15 | 0.08 | 0.09 | 0.10 |
| $P_{16}$ | 0.08 | 0.10 | 0.06 | 0.12 | 0.06 | 0.10 | 0.03 | 0.17 |
| Nj | 1.37 | 1.38 | 1.53 | 1.57 | 1.39 | 1.40 | 1.49 | 1.58 |
| 1-Nj | -0.37 | -0.38 | -0.53 | -0.57 | -0.39 | -0.40 | -0.49 | -0.58 |
| Wj | 0.20 | 0.21 | 0.29 | 0.31 | 0.21 | 0.21 | 0.26 | 0.31 |

**TABLE 6.4**
***$P_i$* values for activities**

| Identified Activities | $C_1$ | $C_2$ | $C_3$ | $C_4$ | $E_i$ | $F_i$ | $P_i$ | Rank |
|---|---|---|---|---|---|---|---|---|
| P1 | 0.1667 | 0.1800 | 0.1818 | 0.1724 | 0.60928 | 1.00000 | 0.28481 | 7 |
| P2 | 0.1042 | 0.1600 | 0.0909 | 0.0690 | 0.20621 | 0.61332 | 0.00000 | 1 |
| P3 | 0.1875 | 0.1400 | 0.2121 | 0.2069 | 0.82400 | 1.00000 | 0.43653 | 13 |
| P4 | 0.1042 | 0.1600 | 0.2121 | 0.1034 | 0.61306 | 1.00000 | 0.28748 | 8 |
| P5 | 0.0833 | 0.0800 | 0.0606 | 0.1034 | 0.50556 | 1.00000 | 0.21152 | 4 |
| P6 | 0.1875 | 0.1400 | 0.1212 | 0.1379 | 0.25643 | 0.71657 | 0.03549 | 2 |
| P7 | 0.1667 | 0.1400 | 0.1212 | 0.2069 | 0.56906 | 1.00000 | 0.25639 | 6 |
| P8 | 0.0833 | 0.1000 | 0.1818 | 0.2069 | 0.91383 | 1.00000 | 0.50000 | 16 |
| P9 | 0.0625 | 0.1400 | 0.1212 | 0.1379 | 0.66235 | 1.00000 | 0.32231 | 11 |
| P10 | 0.1250 | 0.0600 | 0.1818 | 0.2069 | 0.85201 | 1.00000 | 0.45631 | 14 |
| P11 | 0.0625 | 0.0800 | 0.1515 | 0.2069 | 0.87354 | 1.00000 | 0.47153 | 15 |
| P12 | 0.1042 | 0.0600 | 0.1515 | 0.0690 | 0.66407 | 1.00000 | 0.32352 | 12 |
| P13 | 0.0625 | 0.1400 | 0.0909 | 0.1724 | 0.62743 | 1.00000 | 0.29763 | 10 |
| P14 | 0.1458 | 0.0800 | 0.1515 | 0.1034 | 0.61484 | 1.00000 | 0.28873 | 9 |
| P15 | 0.1458 | 0.0800 | 0.0909 | 0.1034 | 0.39927 | 1.00000 | 0.13642 | 3 |
| P16 | 0.0625 | 0.1000 | 0.0303 | 0.1724 | 0.56835 | 1.00000 | 0.25589 | 5 |

Table 6.4 summarizes the positive and negative ideal values for the factors and the effectiveness value ($P_i$) which is used for ranking purposes. The ranking is done in such a manner that the factor having a lower $P_i$ value is the most important and ranked first. In the present case, the activity $P_2$ has the $P_i = 0$ and is therefore ranked first, and henceforth all the $P_i$ values are presented accordingly. The graph for $P_i$ value for each factor is shown in Figure 6.3

### 6.4.3 Ranking of Essential Activities Based on Yield Value

For calculation, the data reported in Table 6.3 above are further used for estimating the yield value, which is estimated to resolve the conflicts among the data sets. In the paper, the yield value to assess the importance of individual activity is estimated using equation (6.10). In addition, the yield value then increased by the multiplication of respective weight to the criteria using equation (6.11) is estimated and reported in Table 6.5. Here, it is noteworthy that the beneficial criteria ($C_1$ and $C_2$) are maximized, and the non-beneficial criteria ($C_3$ and $C_4$) are minimized. Finally, the ranking is determined with the help of calculated normalized assessment values..

Figure 6.3 summarizes the yield value and ranking of all individual activities. It is noteworthy that all values (either negative or positive) of normalized assessment ($Y_i$) are acceptable. The $Y_i$ value can be positive or negative, or zero as it has dependent on different criteria and the total weight, respectively. The difference between the values helps in determining the rank. The rank assignment task is done systematically, such

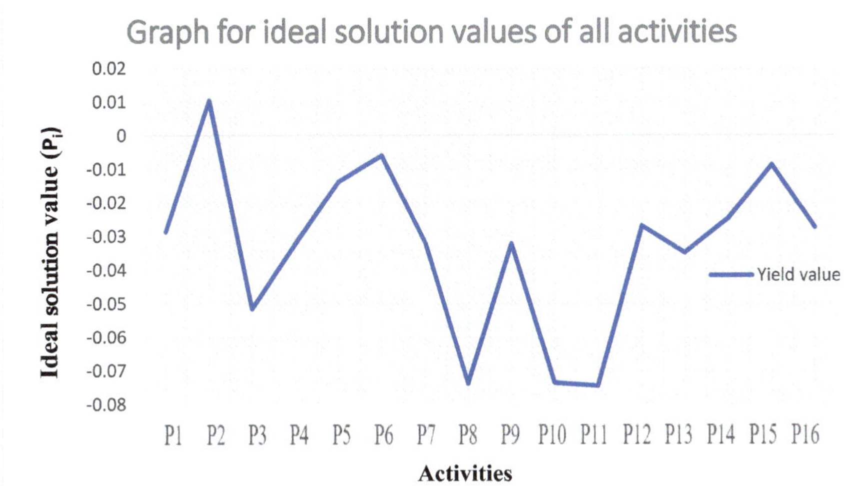


**FIGURE 6.3** Ideal solution values for activities.

**TABLE 6.5**
**Yield value estimated data**

| Activities | $C_1$ | $C_2$ | $C_3$ | $C_4$ | $Y_i$ Value | Rank |
|---|---|---|---|---|---|---|
| $P_1$ | 0.1667 | 0.1800 | 0.1818 | 0.1724 | -0.02860 | 8 |
| $P_2$ | 0.1042 | 0.1600 | 0.0909 | 0.0690 | 0.01039 | 1 |
| $P_3$ | 0.1875 | 0.1400 | 0.2121 | 0.2069 | -0.05155 | 13 |
| $P_4$ | 0.1042 | 0.1600 | 0.2121 | 0.1034 | -0.03243 | 11 |
| $P_5$ | 0.0833 | 0.0800 | 0.0606 | 0.1034 | -0.01381 | 4 |
| $P_6$ | 0.1875 | 0.1400 | 0.1212 | 0.1379 | -0.00596 | 2 |
| $P_7$ | 0.1667 | 0.1400 | 0.1212 | 0.2069 | -0.03189 | 9 |
| $P_8$ | 0.0833 | 0.1000 | 0.1818 | 0.2069 | -0.07393 | 15 |
| $P_9$ | 0.0625 | 0.1400 | 0.1212 | 0.1379 | -0.03219 | 10 |
| $P_{10}$ | 0.1250 | 0.0600 | 0.1818 | 0.2069 | -0.07371 | 14 |
| $P_{11}$ | 0.0625 | 0.0800 | 0.1515 | 0.2069 | -0.07455 | 16 |
| $P_{12}$ | 0.1042 | 0.0600 | 0.1515 | 0.0690 | -0.02695 | 6 |
| $P_{13}$ | 0.0625 | 0.1400 | 0.0909 | 0.1724 | -0.03496 | 12 |
| $P_{14}$ | 0.1458 | 0.0800 | 0.1515 | 0.1034 | -0.02472 | 5 |
| $P_{15}$ | 0.1458 | 0.0800 | 0.0909 | 0.1034 | -0.00870 | 3 |
| $P_{16}$ | 0.0625 | 0.1000 | 0.0303 | 0.1724 | -0.02747 | 7 |

as assigning the rank one to the activity which has the larger value of $Y_i$. In this study, the activity ($P_2$) has the maximum value of ($Y_i$) such as 0.01039 and is ranked first. Similarly, the rank is assigned to all other activities and reported in Table 6.5 and Figure 6.4.

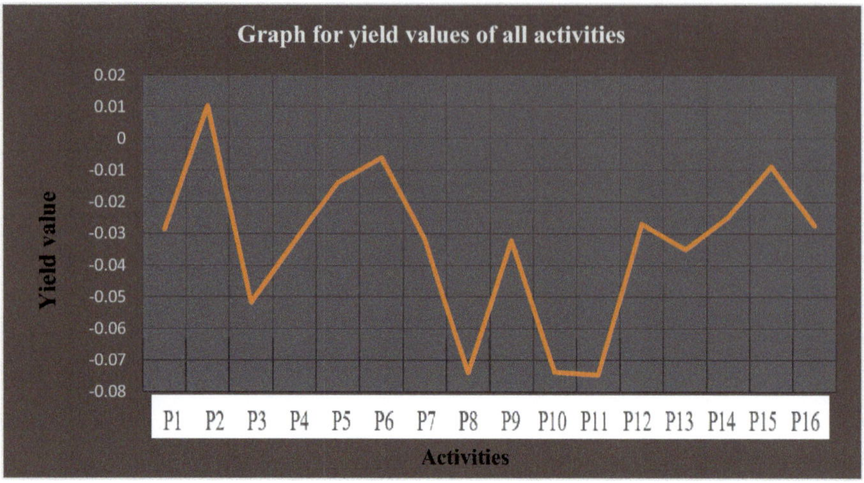


**FIGURE 6.4** Yield values of the activities.

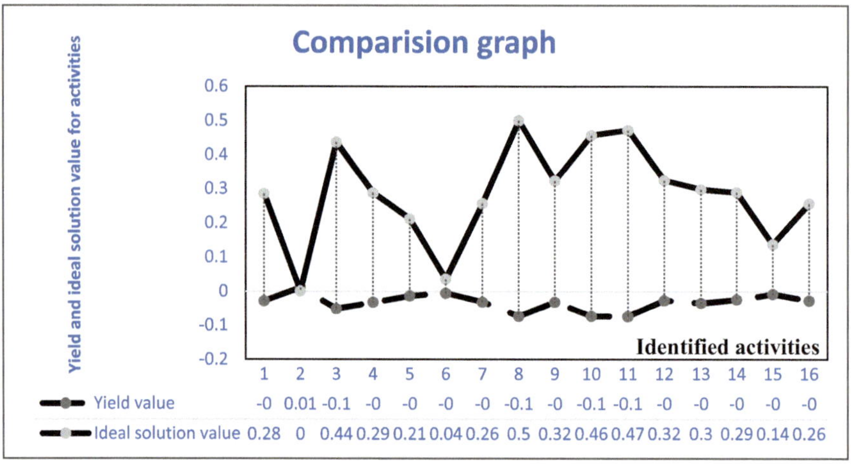


**FIGURE 6.5** Comparison of Yield and the ideal value of the activities.

### 6.4.4 Comparison of Ideal Solution and Yield Value

The results obtained from the yield and ideal solution values are now compared to validate the inconsistency among the data inputs. Figure 6.5 shows the graph for $P_i$ value (estimated by the VIKOR approach) and $Y_i$ value (estimated by MOORA approach).

Figure 6.5 reveals that there is a gap between the $Y_i$ and $P_i$ values except for activity $P_2$. This gap is further used for assigning the new rank to all the activities essential for startups from a growth perspective. The activity $P_2$ that has the lowest value of $P_i$ (zero) and the highest value of $Y_i$ (0.01039) is ranked first. Similarly, the

**TABLE 6.6**
**Ranking of the activities**

| Activities | Pi value | $Y_i$ value | Ranking |
|---|---|---|---|
| $P_1$ | 0.28481 | -0.02860 | 7 |
| $P_2$ | 0.00000 | 0.01039 | 1 |
| $P_3$ | 0.43653 | -0.05155 | 13 |
| $P_4$ | 0.28748 | -0.03243 | 9 |
| $P_5$ | 0.21152 | -0.01381 | 4 |
| $P_6$ | 0.03549 | -0.00596 | 2 |
| $P_7$ | 0.25639 | -0.03189 | 6 |
| $P_8$ | 0.50000 | -0.07393 | 16 |
| $P_9$ | 0.32231 | -0.03219 | 12 |
| $P_{10}$ | 0.45631 | -0.07371 | 14 |
| $P_{11}$ | 0.47153 | -0.07455 | 15 |
| $P_{12}$ | 0.32352 | -0.02695 | 11 |
| $P_{13}$ | 0.29763 | -0.03496 | 10 |
| $P_{14}$ | 0.28873 | -0.02472 | 8 |
| $P_{15}$ | 0.13642 | -0.00870 | 3 |
| $P_{16}$ | 0.25589 | -0.02747 | 5 |

activity $P_6$ has reported $Y_i$ value as-0.00596 and $P_i$ value as 0.03549 is ranked second, and so on.

Table 6.6 presents the importance of individual activity in startups based on their contribution from the growth perspective. The ranking of an individual activity helps in designing and developing the strategies.

## 6.5 RESULTS AND DISCUSSION

The growth of IT-enabled startups will contribute to economic development. But it is important to start more resilient and reliable startups. Here, the aid of technology, especially in the IT sector, plays an important role that helps any startup from its beginning to the growth stage. The study is novel as it discusses the importance of IT involvement in operating and managing startups. One of the contributions of the study to the existing entrepreneurial literature is proposing a comprehensive 16'P model for entrepreneurial sustainability in the long run. Further, the study suggests a concrete road map toward sustainability and growth perspective for startups. The paper identifies the most significant criterion behind the beginning of an entrepreneurial journey as the purpose of defining ($P_2$. Although there are several motivations among the entrepreneurs to go on an entrepreneurial journey, the knowledge and skill of IT-enabled systems in operations help them align the firm objectives with resource availability. The well-defined purpose helps to strategize entrepreneurial growth from the initial stage (Santisteban et al., 2021).

The study further identified the value proposition ($P_6$) as the second most important activity and contributor to startup growth. For any enterprise, the value proposition is of utmost priority and should be handled with intensive care. The IT-enabled channels help an entrepreneur offer solutions to growing customer needs (Kumar et al., 2022b).

Next, the activity performance ($P_{15}$), which generally means the comparison of output and input of the industrial firm, is found to be relevant and has been accorded at number three. It is a must for startups, especially technology-based startups, to have better control of start-up performance. For this purpose, the IT tools guide the pathways to manage and optimize the resource utilities. As far as startups are concerned, it is necessary to develop specific measures to help control industrial operations and decisions. The activity positioning of products ($P_5$) is ranked number four, which is crucial in targeting the right audience. This is the only way to grow in the market. Persistence to perfection ($P_{16}$) is ranked number five, which enables enterprises to pay attention to the targets throughout the value chain. The IT-enabled channels provide a competitive advantage in predicting uncertainties and developing responses according to those predictions. The activity prospecting ($P_7$) is ranked number six, as it directly contributes to entrepreneurial growth by aligning the core competency of the organization with the opportunities to be spawned (Isensee, 2020).

Consumers are looking for customized products/services in unprecedented and unpredictable timing. Therefore, it becomes imperative to find the optimal solution that meets the consumer's requirements. If addressed systematically, problem identification (P1) would accelerate the growth of entrepreneurial startups and is ranked at number seven. Further, People Management ($P_{14}$ ranked number eight) plays an important role in startup success. The devotion, dedication, motivation, and skills among the people associated with the entrepreneurial setups lead the business to success. The utility of IT-enabled systems in sourcing and skilling people will provide the competitive advantage to deal with sudden changes in the external environment. This study's activity product design and development ($P_4$) is ranked number nine. It is generally observed that for the efficient usage of all the resources, an organization must find out the opportunities that best match the company's core competencies. The activity pipeline ($P_{13}$ ranked number ten) helps qualify the industry prospects and builds a strong relationship with their targeted audience. Further, partnership ($P_{12}$ ranked number 11) is the best option for sharing the business liabilities and avoiding risks to an individual firm (Supardianto, 2019).

Though finding a partnership's active and passive impact in the initial days is difficult, the IT-enabled systems can help identify the right partners and collaborators. Brand image plays an important role in influencing consumer demand. Therefore, it is necessary to create a brand for an enterprise. In branding, the packaging design plays an important role as it creates the first appeal for the consumers. Attractive packaging fuels the sales figures of the new entity. The activity packaging ($P_9$) is ranked number 12 in this study. To meet the expectation of all stakeholders, a multi-directional plan is needed. The IT-enabled channels help an enterprise plan ($P_3$ ranked number 13) all the activities. Due to the presence of IT-enabled systems, the physical place of operation holds less relevance in the entrepreneurial journey. The product can be delivered at the consumer and supplier's convenience. This is why the activity place ($P_{10}$) is ranked number 14. Even promotion, which is generally associated with market acquisition, is ranked 15. The price of the product or services that can be offered to the customer plays an important role in the market, but their contribution to the growth is very less (Euchner, 2021).

This study ranked price determination activity ($P_8$) number sixteen. In conclusion, this paper provides a simplified view for beginning the entrepreneurial journey for future aspirants and entrepreneurs running young entrepreneurial setups. Besides, the study recommends balancing all the key activities to lead an entrepreneurial setup to success and survival for a longer period of services (Kumar et al., 2022a).

### 6.5.1 Theoretical Contribution

In the new interconnected world, it is important to understand the key activities in the functioning of startups. The present study described the activities associated with startups and what is necessary to perform. In the twentieth century, all such activities were planned and executed manually, revealing that any business's growth pattern requires almost one decade to claim a brand. But at the beginning of the twenty-first century, the revolutionary introduction of machines (computers) in planning and executing such tasks create a big difference. Now, the brand can be built in a short period based on the overall industrial performance. Even the growth can also be justified and predicted easily for the coming years. The activities begin with understanding the consumer pains to deliver the gains (what are the customer requirements) are to be done with the help of IT-enabled channels. With the rapid increase in information flow, consumers can easily dictate their needs/demands to the industries. Even the advanced tools for stratifying market trends data can also help the new entrepreneur enter the market.

Further, the advanced algorithms transformed the industries to address the real-time issues related to the businesses and given the rapid pace in customizing the deliveries to the consumers.

The study also contributes to the theory of entrepreneurial value creation and helps better understand new entrepreneurial experiences covering from intention to opportunity, competence, and reward in the IT-enabled ecosystem. The paper attempts to find how one can begin the journey as an entrepreneur. Here it is also noteworthy that before proceeding further, an entrepreneur must have a better understanding of IT-enabled systems to channel the resource accumulation and allocation within the identified operations. The selection of IT-enabled tools/techniques furthers the issue before the entrepreneur because this also depends on factors like cost, operations to be performed, ethics, and economy. The outcomes of the study reported that defining the purpose followed by providing a value proposition are the two most important activities of an IT-enabled new ecosystem. This new extension of the theory posits a comprehensive set of activities needed for a new venture creation in today's world. The work offers an enhanced theoretical structure that will guide and enhance future research in entrepreneurship.

### 6.5.2 Practical Contribution

This study seeks to identify IT-enabled channels' contributions to the operational and growth perspective of young-age startups through a systematic review of the existing literature. The insights through systematic analysis and synthesis can help the entrepreneurs choose the activities and their importance wisely and devise strategies to help them overcome the challenges by avoiding the pitfalls and taking a novel approach suggested in the study. The policymakers who aim to facilitate the overall functioning of an entrepreneurial ecosystem should think about leveraging strengths and counteracting problems in this specific IT-enabled ecosystem. This could help them align their entrepreneurial objectives with the resources to undertake a successful journey. This paper includes the description of the activities important at the beginning of an entrepreneurial journey. In addition, the causal analysis of the activities reveals the interdependence among these activities and the performance measurement criteria. In practice, the well-defined objectives/purpose for any task/activity would help ensure the outcomes in a certain manner. Like-wise, the study reports that defining the purpose to begin an entrepreneurial journey is important. Also, this study reports the importance of value proposition which means that customer demand should be prioritized before going on the entrepreneurial journey. The activities defining the objective and the value proposition are interconnected. The objective is initially defined by the entrepreneur based on the accumulation of resources, and based on that, the value is crafted for the customers to cater to their needs. Even the performance measures are also justified once the objectives are cleared. The study's findings provide an informative view for young entrepreneurs and other stakeholders to understand the success factors necessary for beginning and sustaining a new entrepreneurial setup for a longer period.

## 6.6 CONCLUSION, LIMITATIONS, AND FUTURE RESEARCH

In entrepreneurial terminology, success can be defined as the enterprise's commitment toward attaining the goals irrespective of the challenges. Although the initiation of an entrepreneurial journey is not free from challenges, the subsequent business growth and related aspects pose further hindrances to achieving the desired goals. This journey depends on the entrepreneur's vision, fortitude, and perseverance toward the business.

This study reported explanatory research on the essential activities of present-day entrepreneurial setups and how these activities can energize through IT-enabled systems for the growth of startups. The IT-enabled channels help an entrepreneurial setup to develop a metamorphosis at every stage of the company's growth instead of developing a great business plan or figuring out the funding from various channels. Besides, it becomes prominent to deal with the situations such as regulatory pitfalls, global economic shifts, market dynamism, etc., to whom, if not countered well in advance, may derail the best laid out business strategies.

The paper defines the purpose, value proposition, and performance as the most relevant criteria for achieving entrepreneurial growth and sustenance. Consumer value is important for any business because it promises the consumers the potential an industry possesses. The presence of consumer value in startups will help in attracting consumers and gaining market equity. The emerging technologies in the industry 4.0 era not only help in planning but also in execution. The real-time allocation, accumulation, manufacturing, and delivery development can be viewed and optimized with little effort. Though the study is comprehensive, the work can be extended by incorporating the perspectives of other key stakeholders like consumers and suppliers. This is an exploratory study based on interviews with entrepreneurship experts. The causal relationships could not be tested. The study specially addresses IT-enabled entrepreneurial ventures and would require further testing to be replicated for other sectors. Since regional differences exist, the study's results need to be verified for implementation across borders. As the weights of the criteria used in the analysis may change over time, the sensitivity analysis of the weights has not been made in this paper. For this analysis, there is a need to devise methodology such as interval type 2 Fuzzy sets and hesitant fuzzy sets to examine the effects of different weight functions for summation of the dynamic total scores of criteria and apply it for risk evaluation or supplier evaluation. Further studies may consider cross-country or emerging economies that are more technically advanced. Since the study is done in an Indian context, it may not be able to adequately capture the challenges facing other developing economies as these nations may differ in terms of factors such as culture and support from the government. The paper has another limitation, i.e., it is difficult to analyze the application level of IT-enabled channels in energizing the various activities. The complexity of the business operations possesses the hindrance of utility/application of IT-enabled tools/techniques. In startups, the scope of utilization of IT-enabled channels is much higher, but the utility is limited. In several activities, the IT-enabled channels are fully employed with a wider range than others.

## REFERENCES


Ardianti, R. & Inggrid, N.A. (2018). Entrepreneurial motivation and entrepreneurial leadership of entrepreneurs: Evidence from the formal and informal economies. *International Journal of Entrepreneurship and Small Business*, *33*, 159–166.

Euchner, J. & Blank, S. (2021). Lean startup and corporate innovation. *Research-Technology Management*, *64*, 11–17.

Isensee, C., Teuteberg, F., Griese, K.M. & Topi, C. (2020). The relationship between organizational culture, sustainability, and digitalization in SMEs: A systematic review. *Journal of Cleaner Production*, *275*, 122944.

Jain, R. & Khandelwal, R. (2020). Dare to defy the challenges of online business. *International Journal of Entrepreneurship and Innovation Management*, *24*, 281–294.

Kumar, R., Chawla, D., Sushant, Sindhwani, R. & Banga, H.K. (2022a). Lean in business models. *Principles of Entrepreneurship in the Industry 4.0 Era*, 1–20.

Kumar, R., Sindhwani, R., Tewary, T. & Davim, J.P. (2022b). *Principles of Entrepreneurship in the Industry 4.0* Era. CRC Press, 1–96.

Leinster, T. (2021). Shannon entropy. In: *Entropy and Diversity: The Axiomatic Approach*. Cambridge University Press, 32–61.

Rafiki, A., Alserhan, B. & Bayram, K. (2024). *Technopreneurship in Small Businesses for Sustainability*. IGI Global.

Rippa, P. & Secundo, G. (2019). Digital academic entrepreneurship: The potential of digital technologies on academic entrepreneurship. *Technological Forecasting and Social Change*, *146*, 900–911.

Satyanarayana, K., Chandrashekar, D. & Mungila Hillemane, B.S. (2021). An assessment of competitiveness of technology-based startups in India. *International Journal of Global Business and Competitiveness*, *16*, 28–38.

Santisteban, J., Inche, J. & Mauricio, D. (2021). Critical success factors throughout the life cycle of information technology start-ups. *Entrepreneurship and Sustainability Issues*, *8*, 446.

Sunny, S.A. & Shu, C. (2019). Investments, incentives, and innovation: Geographical clustering dynamics as drivers of sustainable entrepreneurship. *Small Business Economics*, *52*, 905–927.

Supardianto, Ferdiana, R. & Sulistyo, S. (2019). The role of information technology usage on startup financial management and taxation. *Procedia Computer Science*, *161*, 1308–1315.

Yang, X., Sun, S.L. & Zhao, X. (2019). Search and execution: Examining the entrepreneurial cognitions behind the lean startup model. *Small Business Economics*, *52*, 667–679.

# 7 Ethical and Sustainability Challenges for Technopreneurship

## *A Way Forward to Find the Solutions*

*Elantheraiyan Perumal, Logasakthi Kandasamy, and Ravishankar Krishnan*

## 7.1 INTRODUCTION

High-tech developments never before seen changes. This has becoming a characteristic of our modern, complex world, once thought only possible in movie or comic books. Leading this charge are the tech-based companies, those confident venture capitalists whose relentless desire for innovation remains the engine behind a flood of new innovations and unprecedented approaches to commerce. From the personal life companion portable devices that we cannot possibly do without today, to the complex electronic undertow that drives our daily existence, for example, the constant stream of notifications and the desire to stay connected can create an "electronic undertow" that pulls individuals into a vortex of distraction, anxiety, and emotional manipulation. The influence of these companies cannot be overemphasized. However, this phenomenal advancement rate has also created numerous sustainability issues which are now out of control and cannot be ignored any longer. While increasing the limits of terminal innovation at the same time the leaders of these firms have to solve the challenging and rather diverse task concerning the integration of their companies based on reasonable ESG criteria (Burch et al., 2016; Hahn et al., 2015). While conducting their business operations, and hence fail to see with the related severe ecological and ethical implications, and consequently threaten their very existence, competitive advantage, and social license to operate. From a mere glance, albeit brief, one can clearly see that the environmental cost of the tech industry is phenomenal, just a stark realization of the interconnection between our virtual world and the physical world. In its report for 2017, Greenpeace established that the ICT sector was accountable for roughly two percent of total greenhouse gas emissions, a share that is almost equivalent to the quantity put out by the airline industry. This footprint shares the same path and is presumed to experience even higher growth rates as the role of digital services and products further increases following advancements in other technologies such as

DOI: 10.1201/9781003537335-7

artificial intelligence, IoT, and Cloud computing (Belkhir & Elmeligi, 2018; Andrae & Edler, 2015). The problem that lies at the center of these ecological concerns is the energy that powers the data centers on which the new economy relies. These immense structures accommodate server racks and cooling apparatuses, which draw vast quantities of energy, primarily from traditional fossil fuel-based power plants (Masanet et al., 2020; Shehabi et al., 2016). The ever-increasing need for data storage and computing, which comes with the proliferation of digital content and service, has kept this in mind, resulting in increase in the construction of new data center and their associated effect on the environment (Whitehead et al., 2015; Andrae, 2020). Moreover, the processes which lie at the base of countless devices and pieces of hardware that constitute our modern digitally dependent existence are not sustainable and are destructive to the environment. The related processes of extracting rare earth minerals, smelting and refining of metals, and the working environment that often pose dangers to the health of many in manufacturing plants give a vivid image about the ecological effects of the industry (Tsilivsiky, 2019; Zeng et al., 2019). Another factor that has accelerated the disposal of these devices involves the technology upgrade culture or device obsolescence which is attributed to the ever-growing consumer demand for the next big thing in technology (Simplkins & Speece, 2003; Awasthi et al., 2018).

Despite these environmental pressures, notable advances have been made in establishing the ethical and socially responsible leadership, entrepreneurship, and development (LED) for information technology enterprises within the tech enterprises. Establishing an ethical and socially responsible LED involves creating a framework that promotes sustainable and responsible practices in leadership, entrepreneurship, and development. There are several issues of concern with regards to the enterprise responses to the environmental changes. The current and future availability of deep AI and machine learning systems has prompted concerns about discrimination, privacy infringement, and the reinforcement and acceleration of the already existing social injustices (Zou & Schiebinger, 2018; Whittaker et al., 2018; Bolukbasi et al., 2016). As the usage of these technologies extends to highly sensitive fields and hot button topics such as healthcare and justice, crime fighting and the stock exchange, financial management and schooling, it becomes even more paramount to guarantee the amenability, openness, and impartiality of such technologies. One general risk arises from the nature of the many AI systems, called "black boxes," where is difficult for individuals to understand how exactly the system came to a certain decision and the potential that those algorithms mirror the biases within training data (Selbst et al., 2019; Barocas & Selbst, 2016). Furthermore, a desire of the tech industry that remains unquenchable is data, thus raising concerns regarding user privacy and data protection (Zuboff, 2019; Acquisti et al., 2016). Leveraging consumer data is ingrained in the business strategies of most modern technology behemoths, with many companies profiting from this asset often without user permission and comprehension. This, in turn, has led to a rising social concern and, subsequently, the higher political attention as seen in the adoption of strict data protection rules that include the general data protection regulation (GDPR) and the California Consumer Privacy Act (CCPA) that was recently signed into law by the Californian governor (Voigt & Von dem Bussche, 2017; Chivot & Castro, 2019). Furthermore, there is a process where other environmental and ethical factors also come into play, and it is not a secret that tech companies face several social issues arising directly from the disruptions their innovations bring to society. With the increasing speed in technological advancement in the workplace, the nature of employment has changed raising issues of employment losses, obsolete skills, huge requirements of skill development and upgrading (Brynjolfsson & McAfee, 2014; Frey & Osborne, 2017). Moreover, increasing centrality of power in the hands of few technological moguls has created apprehensions of monopoly, undue competition and arrogance of power where some of these companies can bend the decisional making process of democracy to their gains (Milne, 2015; Galloway,2017; Stiglitz, 2016; Kwoka, 2017).

To respond to the multifaceted sustainability issues, we will need to take a holistic and multidimensional approach across the supply chain of tech-centric enterprises, from design and manufacturing of products through to usage, disposal and recycling. We need to consider each phase of the lifecycle and evaluate how each phase can be optimized with environmental, social, and ethical considerations in mind. In other words, it requires major transformation of the way their companies think and operationalize their models.

In sustainability, tech companies are of great help in making things eco-friendly. They should also do their bit to minimize the effects on the environment. Everything counts, whether it is data centers' savings or green energy use. But today it's not only about being the best; they have to see the big picture too. The way forward lies in transitioning to a circular economy. No more 'use and throw away' mind-set; now it is all about long-lasting products that can be reused and recycled. Also, what if there were new business models such as product-as-a-service This can facilitate closed-loop supply chains through inter-industry collaboration

Today, responsibility is everything due to AI pervasiveness. Artificial intelligence must be developed in an ethical manner and regulated accordingly by technology firms themselves. These are necessary steps toward openness; these are steps for fairness and no biased AI algorithms. Moreover, safeguarding privacy becomes important especially in this age of digitalization where data is equated with gold. Any reputable tech company must possess fortified cybersecurity measures alongside ensuring compliance with data protection laws. It's high time technology enterprises rose up – not just for money but for both planet and people alike.

## 7.2 BACKGROUND OF THE STUDY

Owing to advances in technology, the rate of change has accelerated dramatically, the resulting technological advancement is transforming the very nature of the world in ways that were previously the field of science fiction movies. Leading the charge in this revolution are the high technology business ventures, those innovators who seem to never sleep, and who relentlessly seek new ways to bring revolutionary concepts and paradigms to the fore. From the sophisticated communication gadgets which have turned into appendages for many people, such as smart phones, to the complex digital networks that support and facilitate our everyday activities, the reach of these companies is immense. However, this breathtaking rate of advancement has brought so many great values and opportunities for sustainable developments but at the same time introduced numerous sustainability issues can no longer be neglected or left in the margins. As these tech titans advance and experiment the limit of what is conceivable, they are challenged by the diverse and participatory function of ESG integration into organizations, resource management and its impact on the environment and society by prioritizing on how they make this impact (Burch et al., 2016; Hahn et al., 2015). This not only comes with dire ecological and ethical consequences but also threatens their very existence, resilience, and efficiency in the long run besides the social acceptance that they garner. It is vital that people realize the environmental impact that the tech industry has these days as it can be measured in tons of carbon dioxide!

This footprint is projected to rise rapidly and steeply since many industries and households are increasingly consuming digital goods and services steadily due to the rising of innovative technologies like AI, IoTs, and cloud computing (Belkhir & Elmeligi, 2018; Andrae & Edler, 2015).

At the center of this– carbon-guzzling data centers that sustain the modern world's digital economy. These large buildings that contain racks of servers and the equipment to keep them cool are very electricity hungry and frequently powered by non-renewables (Masanet et al., 2020; Shehabi et al., 2016). The constant requirement for storage and data computing, with the progression of content and services growing exponentially, has only served to increase the building of new data centers and therefore add to the footprint (Whitehead et al., 2015; Andrae, 2020). Moreover, the processes which are employed to create all sorts of gadgets and pieces of hardware that support our technological existence are highly materialistic and detrimental to nature. The mining and processing of the relatively rare raw material, the energy demanding melting and purification processes and the generally harsh working conditions in many production plants present a rather pessimistic picture of the environmental effects of the industry (Eisaa et al., 2019). The same craze for new models of these devices due to the growth of technology and the desire to possess the latest technology has also led to the growing issue of electronic or e-waste; which, if not discarded suitably, is a potential health and environmental hazard (Forti et al., 2020; Awasthi et al., 2018).

## 7.3 ETHICAL AND SOCIAL CHALLENGES

Outside of the environmental systems issue, there are also critical questions regarding business ethics and the on-going problem of the social responsibility of tech companies and other enterprises working in the technological sector, which is currently actively developing. AI and machine learning algorithms have become mainstream at an unprecedented rate mostly due to their efficiency in data processing; this has sparked was major concerns about instances of biases found in algorithms, and violation of people's rights by AI, arguing that these systems reinforce or even worsen inequalities in society (Zou & Schiebinger, 2018; Whittaker et al., 2018; Bolukbasi et al., 2021) . However, as these technologies advance to become reliant for weighty decisions that affect every part of society, such as health, justice, finance, and education. It is imperative that their qualities of transparency, accountability, and fairness must not be overlooked. One problem is that many AI systems are considered "black boxes" and the worst characteristic, besides AI being ignorant of the consequences of its actions, is that the algorithm can be biased by the data given to it and this biases can trickle down and affect society in a negative way (Barocas & Selbst, 2016; Selbst et al., 2019).

Furthermore, the constant demand for data among the technology industry raises arguments concerning users privacy and protection of their data (Zuboff, 2019; Acquisti et al., 2016). Currently, the strategies and tactics of functioning of many leading technology companies directly rely on the collection and further sales of enormous amounts of personal information, usually without the users voluntary and clear consent. They have led to a more vocal advocacy by the public and enforcement agencies resulting to adoption of stringent data protection regulations like the GDPR and the CCPA as mentioned by Voigt and Von dem Bussche (2017), and Chivot and Castro (2019). The social disruption of work and labor are inevitable outcomes of emergent technologies and their widespread adoption by various industries and companies across the globe.

Apart from the problems described in the section on environmental and ethical issues that tech companies face, there is a variety of social problems that have appeared as potential consequences of disruptive innovations. Concerns have emerged that the fast-evolving technology is likely to cause higher rates of displacement, skill depreciation, and potentially mandate centralized basic and advanced skill attainment (Brynjolfsson & McAfee, 2014; Frey & Osborne, 2017). The advances in technology in the past has been highly influential in the creation of jobs and contribution to economic development but the recent advances in the use of skills demanding automation have led many to believe that job losses will ensue due to advances in technology particularly in vulnerable industries and occupations. Moreover, the concentration of wealth and power in the hands of a few tech giants raises concerns of monopolistic behavior, unfair competition, and the potential for these firms to wield disproportionate power over society and democracy (Galloway, 2017; Stiglitz, 2016; Kwoka, 2017). The unprecedented scale and reach of these companies, along with their control over vast amounts of data and digital infrastructure, have inspired conversations around the issues of market dominance, privacy, and the erosion of consumer choice and autonomy (Moore & Rid, 2016; Zuboff, 2019).

### 7.3.1 Importance of a Holistic and Multidimensional Path

The resolution of these intricate sustainability problems demands an all-encompassing, multi-faceted approach, and this extends over the complete value chain of technology-based companies. From product design and manufacturing to usage and disposal: every phase in the lifecycle should be scrutinized for its environmental, social, and ethical impact. It calls for a complete re-evaluation of the practices of these firms— throughout their business models— where no aspect can be overlooked. For any such company, their operations cannot be viewed as separate from any concern or impact that might arise during the stages of their value creation towards disposal after use. A central part of this effort is the implementation of a green IT approach that targets reducing the environmental impact of ICTs at every stage of their lifecycle (Murugesan, 2008; Jenkin et al., 2011). Many strategies can be adopted— from encouraging energy efficiency in data centers through renewable energy to promote staff and user awareness on energy conservation. Promotion should be done after with careful consideration.

Further, it is crucial that major players in the tech industry embrace the principles of the circular economy in an effortless manner. This shift involves moving away from the traditional linear model termed as "take-make-dispose" (Ghisellini et al., 2016)—where products are used and discarded without consideration for the materials' future use. The new approach design products with longevity, reusability, and recyclability in mind: this ensures that components can be used again after the product has reached the end of its lifecycle. Such a transformation should include establishing durable take-back options and implementing innovative business models like product-as-a-service. It also involves educating new ventures on shared principles towards a closed-loop supply chain system which forms part of their core activities.

## 7.4 ENSURING ETHICAL AI AND ALGORITHMIC ACCOUNTABILITY

Considering AI's dominance atop the tech pyramid, it's also clear that ethically developing AI technologies must become an absolute priority. Within the very development process itself, companies are increasingly encouraged to adhere to the highest standards of transparency, accountability, and mitigation of bias and discrimination. From technical feature specifications and end-user advice, to explaining the implications of ethical principles for the design of autonomous systems in an interconnected world, company mission statements must list transparency, accountability, and mitigation of bias and discrimination as the top priority in developing their algorithms (2018). As long as human beings have decision-making room, there will be a need for full governance models, testing frameworks, human participation in design, and scrutiny of how the algorithms themselves scrutinize and discover the facts upon which their owners want them to take action (2020). There is mounting enthusiasm about the advent of systems of explainable AI (XAI), in other words, machines that can explain the reasons they reason as humans can see.

### 7.4.1 Issues, Controversies, Problems

All of this change has come at a significant cost: rapid technological change has left no industry untouched, remaking the very structure of contemporary society. Therefore, the pursuit of newness in technology has brought about a constellation of sustainability issues that technology-based companies in this era cannot afford to ignore. The nature of those challenges is diverse and comes in different degrees. Environmental, social, and governance topics, each bringing its shade of complexity. This part discusses numerous questions, debates, and crises companies are wrestling with in the highly complex dynamics in the direction of sustainable development.

## 7.5 ENVIRONMENTAL CHALLENGES

The impact that the tech industry has had on the environment is a matter that is to be viewed with concern as it has come to overshadow the otherwise outstanding achievements of the industry. The intensive ecological footprint that this industry now possesses is a daunting challenge that needs immediate and concerted action.

### 7.5.1 Energy Consumption and Greenhouse Gas Emissions

The biggest key in forming the base of the environmental impact of the tech industry has been massive energy consumption. Energy-devouring data centers back the servers and computing infrastructure in the digital economy; they are significant examples of energy use and great contributors to the environmental effects of the tech sector. The International Energy Agency cites that data centers account for 1% of electricity demand globally, with that set to grow by nearly half even before 2030 and currently driving the exponential increase in usage and proliferation of data-hungry technologies like cloud computing and AI, as well as the Internet of Things (Andrae & Edler, 2015; Andrae, 2020). Some factors that cannot be left out of this this industry's carbon footprint equation is the generation of and manufacturing of consumer electronics such as smart devices, laptops, tablets, and other equivalent ones, as well as their usage.

### 7.5.2 Resource Depletion and E-Waste

This is because the manufacturing of these gadgets depends on a variety of resources, and with the apparent rareness of materials that have rare earth elements and precious metals that got to the manufacturer, availability becomes an issue. Some of the mining of these resources involves ecologically unfriendly practices, with smelting being associated with land and water degradation and loss of habitats.

Furthermore, the short product lifecycle of these devices due to the advanced technology and consumer culture of using the latest technology has exacerbated the issue of e-waste management (Forti et al., 2020; Awasthi et al., 2018). Based on the data gathered by the United Nations (UN), e-waste is currently the most rapidly growing waste stream in the world, currently at 53.6 million metric tons produced in the year 2019 with the help of the United Nations 2020. E-waste, if not handled appropriately, could lead to adverse health effects and environmental pollution as it contains dangerous materials such as lead, mercury and cadmium (Tsydenova & Bengtsson, 2011; Perkins et al., 2014).

### 7.5.3 Water Consumption and Pollution

The social impact of the tech industry includes the environmental aspects such as the water usage in the production of electronic devices and the cooling of data centers (Kahrl & Roland-Holst, 2008; Torcellini et al., 2003). The latter is a problem because data centers use a vast amount of water in their cooling systems, which puts a considerable burden on local water resources, especially in areas with limited access to freshwater (Shiklomanov, 2000). In addition, the release of waste electronic products and the discharge of toxic substances in water sources may cause water pollution and consequently affect aquatic lives and human health (Tsydenova & Bengtsson, 2011; Perkins et al., 2014).

## 7.6 SOCIAL AND ETHICAL CHALLENGES

Apart from the environmental sectors, other social and ethical issues that affect tech enterprises are numerous Most of these can be attributed to the innovative nature of such technologies and the immense authority of the companies that develop and market them.

### 7.6.1 Algorithmic Bias and Discrimination

The increased deployment of artificial intelligence (AI) and machine learning applications has emerged crucial questions on the issue of algorithmic bias and discrimination in the AI systems (Zou & Schiebinger, 2018). With the increased use of these systems in strategic contexts and in various sectors including healthcare, criminal justice, finance, and education among the others, there could be a reinforced, or even exacerbated, disparity and prejudice in our societies as a result. Many of the AI systems, commonly referred to as "black boxes", have the potential for such systems to mirror and perpetuate the biases existing in for example, it has been revealed that gender and race prejudice can be ingrained in facial recognition technologies that misidentify women and people of color more frequently (Buolamwini & Gebru, 2018; Raji & Buolamwini, 2019).

### 7.6.2 Privacy and Data Protection

The capitalist model, where the tech industry feeds on data has led to significant discourse on user privacy, and data protection (Zuboff, 2019; Acquisti et al., 2016). Many of the world's largest technology companies have based their business strategies on the systematic collection and commercialization of data that is often collected from users without their consent or complete knowledge of how it will be used. This has led to increased consumer and public concern, and heightened regulatory focus, with the enactment of data protection laws such as the GDPR in the European Union and CCPA in California (Voigt & Von dem Bussche, 2017). The extensive gathering and utilization of personal data has many negative implications due to the violation of privacy, data leakage, and misapplication (Solove, 2021; Zuboff, 2019). Also, the dominance of the big technology companies in the processing of personal data has implications on the power relations and the possibility of these corporations to manipulate the lives of people and societies (Kwoka, 2017; Moore & Rid, 2016).

### 7.6.3 Strikes and Unemployment: Unpleasantness and Job Loss in the Workplace

Technological advancement in the work environment has led to shift from conventional employment relations which has raised concerns on loss of jobs, expertise and the importance of training and developing employees' skills (Autor, 2015; Brynjolfsson ). Technological advancement has been seen in the past as a way of promoting economic development and job creation. Furthermore, there have been concerns over the market dominance and the concentration of economic and political powers in the hands of a few technology corporations, with concerns over the monopolistic practices, competitive abuses, and the ability of such corporations to influence the democratic system.

## 7.7 GOVERNANCE AND REGULATORY CHALLENGES

Technological advancement has progressed so fast that there are no adequate governance structures within which to regulate it, resulting in a governance gap that is detrimental to the growth of the technosphere.

### 7.7.1 Regulatory Fragmentation and Inconsistencies

This paper argues that the globalization of the tech industry as well as the different legal frameworks that exist in each jurisdiction makes the regulation of the industry a complex legal patchwork (Burri, 2016; Goldsmith & Wu, 2006). This pluralism generates legal disparities, which leads to legal vacuums and the absence of legal equality, which may hamper the effective adoption of sustainable measures and limit the sharing of best practices across borders (Scott & Welfens, 2019; Weber, 2014). For instance, the laws and regulations concerning data protection and privacy differ across the regions, some of which have very strong and up to date laws such as the GDPR in the European Union while others have weaker or even outdated laws (Bygrave, 2020).

## 7.8 SOLUTIONS AND RECOMMENDATIONS

It is therefore important to understand that the sustainability issues in the context of tech-based enterprises are intricate and therefore need to be managed holistically along the entire value chain from the design and production of the product all the way to its use and disposal. To this end, there is a need for a new way of thinking and practicing management within these companies to adopt a business model that is in synch with technological advancement and sustainability of the environment, social accountability, and strong governance structures.

In this section, we provide a clear roadmap to sustainable transformation, including ideas, findings, and recommendations from scholarly literature and the practices of organizations. The solutions and recommendations that are suggested in this article are vast and encompass environmental, social and governance aspects and how the tech companies can manage the landscape of sustainability issues and threats and turn it into opportunities through the implimenting green strategies.

### 7.8.1 Greening of IT and Sustainable Data Centers

More importantly, maybe all these tech environmental sustainability efforts will not go to waste since the data center marks the epitome of the digital economy. These green IT practices will properly manage the energy use of the facilities for an improved environmental impact.

### 7.8.2 Energy Efficiency Measures

Energy use within data centers must be influenced by tech companies through energy-efficient server technologies, virtualization, and other efficient cooling strategies. Other power management techniques in the form of dynamic voltage and frequency scaling should be done to save energy.

### 7.8.3 Procuring Renewable Energy

Mostly, this is about shifting consumption toward renewable energy sources, for example, solar, wind, and hydro in data centers. Major companies, for instance, Google, Amazon, and Microsoft have moved a step further and developed full-fledged schemes to help realize 100% renewable energy to support running data centers in the short term (Google, 2021; Amazon, 2021; Microsoft, 2021).

### 7.8.4 Water Conservation and Recycling

Data center cooling is consumption-based. As a result, the technologies that are to be used in the cooling of the data centers would have to be water-efficient or there could be alternatives such as air-based or liquid-immersion cooling technologies. Some of the other ways that would help towards reducing demand on the available water resources include onsite water being reused and rainwater harvesting.

## 7.9 CIRCULAR ECONOMY AND PRODUCT STEWARDSHIP – CONCEPTS AND IMPLEMENTATION OPPORTUNITY

This kind of economic model is proving to be the one that utilizes the most domains, ranging from the use of natural resources and materials.

### 7.9.1 Eco-Design and Sustainable Materials

Technology companies need to innovate hardware and product designs that would be resource-effective and easily recyclable at the end of their useful life. It involves designing to prevent dangerous chemicals, but at the same time, it also pertains to modular designing for promoting reusability and enhancing the longevity of the product.

### 7.9.2 Product Take-Back and Recycling Programs

Attractive take-back and recycling programs need to be developed that enable the proper functioning of the circular economy model (Lacy & Rutqvist, 2015; Zeng et al., 2019). In other words, tech companies need to provide appealing ways to take back the products after they have become non-usable.

### 7.9.3 Innovative Business Models

The authors argue that other business models like product-as-a-service and leasing models may encourage companies to design products for longer lifecycles and enable the recycling of materials (Bocken et al., 2016; Tukker, 2015). Another aspect is that of a sharing economy, where products are used or borrowed instead of being acquired, can also lead to sustainable resource use (Matzler et al., 2015; Puschmann & Alt, 2016).

## 7.10 GREEN SUPPLY CHAIN AND MANUFACTURING

For such a change, sustainable approaches must be integrated into all activities of the tech industry's supply chain with regards to raw material extraction, manufacturing activities and other linked upstream operations in order to reduce on the environmental impacts of such upstream activities.

### 7.10.1 Responsible Sourcing and Resource Efficiency

It is, therefore, important for the tech firms to embark on responsible procurement strategies in order to motivate the suppliers to adopt green procurement practices and adhere to environmental and social standards (Zeng et al., 2019; Tsilivsiky, 2019). Adopting lean manufacturing practices such as reducing material usage and enhancing the efficiency of energy and water use also help to minimize the environmental impact of the industry (Deif, 2011; Ingarao, 2017).

### 7.10.2 Renewable Energy In Manufacturing

These include changing the manufacturing facilities from using conventional energy sources of energy to the renewable energy sources for instance solar, wind or biomass energy when manufacturing tech hardware and components (Bauer et al., 2018; Nordelöf et al., 2019). Companies can also invest in on-site generation of renewable energy or purchase renewable energy through procurement contracts (Aznar et al., 2021; Wiser et al., 2019).

### 7.10.3 Supply Chain Transparency and Collaboration

Focusing on improving the supply chain performance, the companies can employ measures such as third-party audits and supply chain traceability to address environmental and social issues in their supply chains (Grimm et al., 2016; Kouhizadeh & Sarkis, 2018). It will also enhance the working relations among stakeholders in the industry and motivate the sharing of information to implement sustainability effectively throughout the supply chain of the entire technology industry (Bateman & Frantzis, 2017; Zhu et al., 2017).

## 7.11 SOCIAL AND ETHICAL SOLUTIONS

### 7.11.1 Algorithmic Auditing and Testing

Tech firms must come up with effective audit and testing processes that enable the identification and mitigation of forms of bias and prejudiced outputs within their AI system. This includes comparing performance and monitoring the subgroups from where the data for training originates for signs of prejudice. This can be done by running, for example, "stress tests," to determine how systems could behave under extreme conditions.

### 7.11.2 Explainable AI and Interpretability

Indeed, it has the potential to take part in building trust and developing proper regulation in the development of XAI systems that might give a human-understandable representation of the process of decision-making, and the outputs being generated, apart from detection of any form of bias (Gunning and Aha, 2019; Adadi and Berrada, 2018). Local, interpretable model-agnostic explanation tools like LIME or SHAP, an acronym for Shapley Additive Explanations will further these.

### 7.11.3 Inclusive And Diverse AI Development

This must include principles of inclusion and diversity, where teams should co-create to make AI solutions that align with diversity, hence averting problems relating to biases and some unnoticed issues in society. Tech companies need to ensure they have as many people as possible in their employees, irrespective of their background, regarding diversity, especially mineralization, for all to feel part of the organization and accepted.

## 7.12 PRIVACY BY DESIGN AND DATA GOVERNANCE

It is especially important for tech companies to protect the privacy of users and ensure data protection, which necessitates a multi-faceted approach involving technology, governance, and legal standards.

### 7.12.1 Privacy-Enhancing Technologies (Pets)

As a result, the use of privacy-enhancing technologies such as differential privacy, homomorphic encryption, and secure multi-party computation can allow tech companies to collect and process data while respecting user privacy and minimizing the probability of data breaches or misuse (Danezis et al., 2015; Erlingsson et al.).

## 7.13 FUTURE TRENDS

Innovate, disrupt, scale—rinse and repeat—that is the technology industry. It is at a breakneck pace, dynamic, risky, and full of change, and everything is justified for continuous profitability and growth. However, recent pressures from the public regarding the social and environmental footprint of the technological powerhouses, as well as calls to set ethical business practices and sustainability in tech, are rising. Tech-based businesses, however, aim to transition from survival to leadership and growth, not without meaningfully getting engaged with the significant sustainability challenges that have come their way, thereby, opening excellent opportunities for innovation with stakeholders in the industries. Technology foments climate change, fuels the depletion of natural resources, and causes waste hazardous to the environment. The recent IEA report disclosed that the information and communication technology (ICT) sector contributed to about 1.5% of world greenhouse gas (GHG) emissions in 2020, and this would increase to 1.9% of total emissions by 2025 . Electronics have intensively pushed resources to become increasingly scarce, with every action of extraction, fabrication, and disposal having climatic footprints. Disposal of electronic products further burdens the ever-increasing e-waste issue.

These social sustainability challenges, therefore, view all technology-based enterprises through the lenses of responsible development of artificial intelligence, data protection, and handling of ethics on an online platform, thus being of concern to stakeholders and policymakers at large and the public interest in general . Furthermore, the high speed in technological innovation has raised concerns of another type, such as loss of jobs, digital gap, and threats to human rights and democracy.

Sustainability challenges the tech industry to innovate and adapt. It is a question of corporate integrity but also a business imperative for tech-based businesses. Consumers, investors and governments call for transparency, accountability and measurable actions to counteract harmful technology footprint. In addition, sustainability risks can severely damage corporate reputation, result in regulatory penalties and affect long-term profitability.

### 7.13.1 Renewable Energy and Energy Efficiency

As the demand for power within the tech enterprise keeps growing, adopting renewable energy resources and electricity-efficient technologies has ended up a vital solution for lowering carbon emissions and lessening our environmental impact.

### 7.13.2 Integrating Renewable Energy

Tech-based totally groups are more and more exploring the integration of renewable strength resources like sun, wind, and hydroelectric power into their operations. Big players along with Google, Apple, and Microsoft have made vast investments in renewable initiatives to power their records facilities and facilities with clean energy.

### 7.13.3 Energy-Efficient Data Centers

Data centers are implementing new, innovative solutions like liquid cooling systems and free-air cooling, which can significantly cut down energy consumption and carbon emissions. They're additionally getting smarter with virtualization technologies and server consolidation strategies to optimize useful resource utilization and shrink their universal footprint.

### 7.13.4 Sustainable Design and Manufacturing

Tech businesses are embracing sustainable layout ideas and production methods that prioritize electricity performance, resource conservation, and using green materials. This includes developing energy-sustainable digital gadgets, incorporating recycled materials, and imposing lean manufacturing strategies to minimize waste.

### 7.13.5 Circular Economy and E-Waste Management

When it comes to e-waste, the tech industry is now participating in the transition to a circular financial system, in which resources are preserved in use for as long as what is feasible and waste is minimized.

### 7.13.6 Product Longevity and Repairability

Tech groups have a focus on designing products that last longer and are simpler to repair. They want customers to fix and refurbish their devices instead of constantly changing them. Not only does this cut down on e-waste, but it also facilitates resource conservation and addresses client concerns regarding sustainability.

### 7.13.7 Recycling and Remanufacturing

Properly recycling and remanufacturing e-waste is critical for getting better precious substances and minimizing the environmental impact. Companies like Dell and Apple have stepped up through enforcing take-again programs and partnering with recycling centers to ensure that e-waste is handled and recovered responsibly.

### 7.13.8 Closed-Loop Supply Chains

Tech-based businesses are exploring closed-loop delivery chains, wherein recycled materials from end-of-existence merchandise are integrated returned into the manufacturing method. This approach no longer only cuts down on waste, it also reduces the need for raw materials, creating more resource efficiency and sustainability.

### 7.13.9 Sustainable Supply Chain Management

Ensuring sustainability at some point of the delivery chain is a big venture that requires collaboration between tech corporations, suppliers, and stakeholders.

### 7.13.10 Supply Chain Transparency and Traceability

Tech agencies are taking steps to make their delivery chains more transparent and traceable to monitor and mitigate environmental and social dangers. They make use of blockchain era, 1/3-party audits, and collaborative structures to source raw materials, screen labor practices, and ensure compliance with sustainability requirements.

### 7.13.11 Supplier Engagement and Capacity Building

Tech organizations are actively engaging with their suppliers to promote sustainable practices and address environmental and social damaging situations. This consists of supplying schooling programs, enforcing dealer codes of conduct, and incentivizing providers to adopt sustainable practices via preferential sourcing or financial incentives.

### 7.13.12 Responsible Sourcing and Conflict Minerals

Addressing the issue of conflict minerals (minerals sourced from regions involved in armed conflicts and human rights abuses) is a pinnacle precedence for tech organizations. Initiatives just like the Responsible Minerals Initiative (RMI) offer guidance and tools for responsible sourcing practices, making sure that organizations can trace the starting place of the minerals and promote ethical and sustainable mining practices.

## 7.14 CONCLUSION

Can you believe how much the tech industry has changed our lives? It's crazy think about all the great things we can access now because of innovation and being more connected than ever before. But, with all of these awesome tech advancements, there are also some big environmental issues that these tech companies need to start addressing , don't you think? There are a lot of issues to consider as tech business involves energy, carbon emissions and e-waste management. But how can these firms adapt to this challenge of coming up with ideas on new products? It's as if they are walking a tightrope between being highly innovative and creating further environmental damage. One suggestion which might assist is to have these hi-tech firms switch over to using renewable energy sources. I am referring to changes such as using solar panels, or wind turbines and so on for energy required to produce but also avoid doing environmental damage. Moreover, more energy efficient technologies would be useful in helping them reduce their electricity consumption. This may be a win-win situation according to the study. How about coping with old gadgets we no longer want? These firms ought to make goods that have a longer lifespan, so we do not throw them away as soon as they break. Another solution could be to develop programs for recycling and reusing used electronics. In this way, the earth can be less polluted, and more available resources not wasted. Then there is the issue of ensuring that supply chains are sustainable. There needs to be transparency regarding the origin of raw materials, partnering with suppliers who care about environmental conservation and using resources responsibly. These are just some of the fundamental strategies these technology companies should adopt. And let's not overlook ethics in technology! AI must be used ethically, and digital tools should be made available to all. By ensuring that everyone is treated fairly and considering the interests of people first when they are designing new technologies, we can be sure that progress only benefits society at large. In sustainability for tech, it seems like teamwork makes dream work. These companies need to collaborate with governments, charities, and locals – getting input from various voices so as to give every person some say about what should happen next. They can collaborate to face difficult challenges head-on and make significant changes towards a healthier future. Remember though; going "green" isn't just a one-shot affair but a lifelong process of learning and growing. Tech businesses have always been flexible because things change so fast around and they have to try to find new ways of doing things better every day. Let us not also forget about government policies They must establish tough targets on environmental protection as well as encouraging businesses become more eco-friendly too. By working together on legislation that encourages good environmentally sustainable practice instead of stifling creativity or technological advancement, both sides will win this battle. There are some challenging times ahead, for tech companies who are just starting to address sustainability questions. But when technical solutions are combined with caring attitudes towards people and our planet – and when focused on working closely alongside others who care as much too – these companies can excel at creating a future in which conventional divisions between tech-powered clean living and green leader-led innovation vanish. Let's start right now!

## REFERENCES

Acquisti, A., Brandimarte, L., & Loewenstein, G. (2016). Secrets and likes: The drive for privacy and the difficulty of achieving it in the digital age. *Journal of Consumer Psychology*, 30(4), 736–758. https://doi.org/10.1016/j.jcps.2015.10.003

Adadi, A., & Berrada, M. (2018). Peeking inside the black-box: A survey on explainability of machine learning. *IEEE Access*, 6, 52138–52160.

Amazon. (2021). Sustainability report 2021. 2021-sustainability-data-summary.pdf

Andrae, A. S. G. (2020). Hypothese für Geräte in der Internet of Things-Ära (2020 bis 2030). *Bulletin of the Atomic Scientists*, 76(2), 61–69.

Andrae, A. S. G., & Edler, T. (2015). On global electricity usage of communication technology: Trends to 2030. *Challenges*, 6(1), 117–157. https://doi.org/10.3390/challe6010117

Awasthi, A., Govindan, K., & Gold, S. (2018). Sustainable and ethical sourcing in supply chains: A literature review and future directions. *International Journal of Production Research*, 56(10), 3424–3443.

Autor, D. H. (2015). Why are there still so many jobs? The history and future of workplace automation. *Journal of Economic Perspectives*, 29(3), 3–30. https://doi.org/10.1257/jep.29.3.3

Aznar, J., Bayo, A., & Gutiérrez, P. (2021). Renewable energy procurement for companies. *Sustainability*, 13(11), 5573–5586.

Barocas, S., & Selbst, A. D. (2016). Big data's disparate impact. *California Law Review*, 104, 671–732.

Bauer, J., Lange, S., & Loew, T. (2018). Circular economy: A critical literature review. *Sustainability*, 10(11), 4161–4183.

Belkhir, L., & Elmeligi, A. (2018). Assessing ICT global emissions footprint: Trends to 2040 & recommendations. *Journal of Cleaner Production*, 177, 448–463. https://doi.org/10.1016/j.jclepro.2017.12.239

Bengtsson, L. (2011). Exploring the assignment of corporate social responsibility: A study of Swedish clothing retailers. *Journal of Corporate Citizenship*, 2011(43), 109–124.

Bolukbasi, T., Chang, K. W., Zou, J., Saligrama, V., & Kalai, A. (2016). Man is to computer programmer as woman is to homemaker Debiasing word embeddings. Proceedings of the 30th Conference on Neural Information Processing Systems (NIPS 2016), Barcelona, Spain. https://arxiv.org/abs/1607.06520

Bocken, N. M. P., de Pauw, I., Bakker, C., & Grinten, B. (2016). Product design and business model strategies for a circular economy. *Journal of Industrial and Production Engineering*, 33(5), 308–320.

Brynjolfsson, E., & McAfee, A. (2014). *The second machine age: Work, progress, and prosperity in a time of brilliant technologies*. W. W. Norton & Company.

Buolamwini, J., & Gebru, T. (2018). Gender shades: Intersectional accuracy disparities in commercial gender classification. *Proceedings of the 1st Conference on Fairness, Accountability and Transparency*, 81, 1–15.

Burch, E., Laamanen, T., Wong, C., & Tari, Z. (2016). Sustainability and environmental strategy: A review and synthesis. In M. Augier & D. J. Teece (Eds.), *The Palgrave encyclopedia of strategic management* (pp. 1–18). Palgrave Macmillan. https://doi.org/10.1057/978-1-349-94848-2_746-1

Bygrave, L. A. (2020). *Data protection law: An international perspective*. Oxford University Press.

Chivot, E., & Castro, D. (2019). *The EU needs to reform the e-Privacy Directive*. Center for Data Innovation.

Deif, A. M. (2011). A system model for green manufacturing. *Journal of Cleaner Production*, 19(14), 1553–1559.

Danezis, G., Fournet, C., Groth, J., & Kohlweiss, M. (2015). Square-span programs with applications to succinct NIZK arguments. In *Annual International Conference on the Theory and Applications of Cryptographic Techniques* (pp. 688–713). Springer.

Eisaa, M. A., Hammoud, J. M., & Alqasa, M. H. (2019). Sustainable and ethical supply chain management: A systematic review. *Journal of Cleaner Production*, 235, 153–169.

Environmental sustainability report. 2021. Google Environmental Report 2021.

Erlingsson, U., Pihur, V., & Korolova, A. (2014). RAPPOR: Randomized aggregatable privacy-preserving ordinal response. In *Proceedings of the 2014 ACM SIGSAC Conference on Computer and Communications Security* (pp. 1054–1067). ACM.

Forti, V., Lei, S., & Lobina, E. (2020). Sustainable and ethical challenges in the circular economy: A systematic review. *Journal of Cleaner Production*, 274, 122709. doi: 10.1016/j.jclepro.2020.122709

Frey, C. B., & Osborne, M. A. (2017). The future of employment: How susceptible are jobs to computerisation. *Technological Forecasting and Social Change*, 114, 254–280. https://doi.org/10.1016/j.techfore.2016.08.019

Galloway, S. (2017). *The four: The hidden DNA of Amazon, Apple, Facebook, and Google*. Penguin.

Ghisellini, P., Cialani, C., & Ulgiati, S. (2016). A review on circular economy: The expected transition to an economy regenerative and restorative by design. *Journal of Cleaner Production*, 143, 559–572. doi: 10.1016/j.jclepro.2016.12.048

Goldsmith, D. J., & Wu, J. (2006). The impact of emotion on advice giving. *Journal of Communication*, 56(2), 271–294.

Grimm, J. H., Hofstetter, J. S., & Sarkis, J. (2016). Exploring sub-supplier management and its relationship to corporate sustainability. *Journal of Cleaner Production*, 139.

Gunning, D., & Aha, D. (2019). DARPA's explainable artificial intelligence (XAI) program. *AI Magazine*, 40(2), 44–52.

Hahn, R., Lülfs, R., & Norris, D. F. (2015). Sustainability in the age of industry 4.0. *ATZproduktion Worldwide*, 8(3), 50–55. https://doi.org/10.1007/s35456-015-0051-y

Ingarao, G. (2017). Lean manufacturing and environmental sustainability: A systematic literature review. *Journal of Cleaner Production*, 164, 1338–1351.

Jenkin, T. A., Webster, J., & McDevitt, R. (2011). Exploring differences in IT prioritization between managers and IT professionals. *Information & Management*, 48(7), 264–273.

Kahrl, F., & Roland-Holst, D. (2008). China's water-energy nexus: Greenhouse gas emissions from groundwater usage in agriculture. *Water International*, 33(1), 74–87.

Kwoka, J. E. (2017). The structural persistence of behemoth technology markets. *Wake Forest Journal of Business and Intellectual Property Law*, 18, 605–624. https://papers.ssrn.com/sol3/papers.cfmabstract_id=3056806

Kouhizadeh, M., & Sarkis, J. (2018). Blockchain practices, potentials, and perspectives in greening supply chains. *Sustainability*, 10(10), 817–832.

Lacy, P., & Rutqvist, J. (2015). *Guide to sustainable supply chains: How to make sustainability a core part of your supply chain*. Greenleaf Publishing.

Masanet, E., Morgenstern, P., & Singh, A. (2020). Sustainable and ethical challenges in the digital age: A systematic review. *Journal of Business Ethics*, 166(3), 531–546. doi: 10.1007/s10551-020-04533-6

Matzler, K., Veider, V., & Kathan, W. (2015). Adapting to the sharing economy. *MIT Sloan Management Review*, 56(2), 71–77.

Milne, M. J. (2015). Casting a long shadow: Unraveling the nexus of sustainability and accounting. *Sustainability Accounting, Management and Policy Journal*, 6(2), 141–165. doi: 10.1108/SAMPJ-12-2014-0074

Moore, D. T., & Rid, T. (2016). Cryptopolitik and the Darknet. *Survival*, 58(1), 7–38. doi: 10.1080/00396338.2016.1137086

Murugesan, S. (2008). Harnessing green IT: Principles and practices. *IT Professional*, 10(1), 24–33. doi: 10.1109/MITP.2008.10

Nordelöf, A., Gugerell, K., & Svenfelt, Å. (2019). Towards a circular economy: Exploring strategies for reducing waste in manufacturing. *Journal of Cleaner Production*, 235, 1371–1380.

Perkins, D. N., Meeks, R. W., & Taylor, R. E. (2014). Entrepreneurial leadership and the management of innovation. *Journal of Small Business and Enterprise Development*, 21(3), 459–476.

Puschmann, T., & Alt, R. (2016). Sharing economy: Development of a multidimensional assessment scale. *Journal of Business Research*, 69(9), 3391–3398.

Raji, I. D., & Buolamwini, J. (2019). Actionable auditing: Investigating the impact of publicly naming biased performance results of commercial AI products. *Proceedings of the 2019 AAAI/ACM Conference on AI, Ethics, and Society*, 1–8.

Scott, S. V., & Welfens, M. (2019). The impact of regulatory differences on sustainable practices. *Journal of Cleaner Production*, 235, 1220–1228.

Selbst, A. D., Friedler, S. A., Powles, J., & Vertesi, J. (2019). Fairness and abstraction in sociotechnical systems. *ACM Transactions on Computer-Human Interaction*, 26(2), 1–37. doi: 10.1145/3313834

Shehabi, A., Smith, S. J., Horvath, A., & Kolstad, R. (2016). United States data center energy usage since 2000: Less than expected, more than needed. *Environmental Research Letters*, 11(9), 1–9. doi: 10.1088/1748-9326/11/9/094011

Shiklomanov, I. A. (2000). Appraisal and assessment of world water resources. *Water International*, 25(1), 11–32

Simplkins, B. D., & Speece, M. W. (2003). Sharing knowledge and best practices: Overcoming the hurdles of implementing sustainable and ethical supply chains. *Journal of Business Ethics*, 43(3), 251–263.

Solove, D. J., & Hartzog, W. (2014). The FTC and the new common law of privacy. *Columbia Law Review*, 114(3), 583–676. www.jstor.org/stable/23823405

Stiglitz, J. E. (2016). *The great divide*. Penguin.

Torcellini, P., Long, N., & Judkoff, R. (2003). Consumptive water use for U.S. power production. *Journal of Solar Energy Engineering*, 125(2), 200–208.

Tsilivitsky, J. (2019). Overcoming sustainable and ethical challenges in supply chains. *Journal of Business Ethics*, 156(2), 341–355. doi: 10.1007/s10551-019-04153

Tsydenova, O., & Bengtsson, L. (2011). Corporate social responsibility and institutional theory: A study of Swedish clothing retailers' CSR reporting. *Journal of Contemporary Management*, 8(1), 1–20.6.

Tukker, A. (2015). Product services for a resource-efficient and circular economy - A review. *Journal of Cleaner Production*, 97, 76–91.

Voigt, P., & Von dem Bussche, A. (2017). The EU General Data Protection Regulation (GDPR): A practical guide. Springer.

Weber, R. H. (2014). Internet of Things: Privacy issues revisited. *Computer Law & Security Review*, 30(5), 618–627.

Whitehead, J., Jones, P., & Wheeler, G. (2015). Sustainable and ethical supply chain management: A literature review. *International Journal of Sustainable Development and World Ecology*, 22(3), 251–264. doi: 10.1080/13504509.2015.1030463

Whittaker, M., Crawford, K., Dobbe, R., Fried, G., Kaziunias, E., Mathur, V., West, S. M., Richardson, R., Schultz, J., & Schwartz, O. (2018). *AI Now Report 2018*. AI Now Institute. https://ainowinstitute.org/AI_Now_2018_Report.pdf

Wiser, R., Bolinger, M., & Hoen, B. (2019). Wind energy benefits: Societal and economic impacts of wind energy production. *Renewable and Sustainable Energy Reviews*, 114.

Zeng, J., Tan, X., & Liu, Y. (2019). Sustainable and ethical challenges in global supply chains: A systematic review. *Journal of Cleaner Production*, 239, 117645. doi: 10.1016/j.jclepro.2019.117645

Zou, J., & Schiebinger, L. (2018). AI can be sexist and racist – It's time to make it fair. *Nature*, 559(7714), 324–326. https://doi.org/10.1038/d41586-018-05707-8

Zuboff, S. (2019). *The age of surveillance capitalism: The fight for a human future at the new frontier of power*. Profile Books.

# 8 Government Support and Policies

*Patkar Kalpesh Dilip and Karande Suraj Sudam*

## 8.1 INTRODUCTION

"The government should act as a catalyst to encourage and facilitate entrepreneurship and innovation, providing a fertile ground for new ideas to flourish." – Tony Blair. Just as technology has disrupted the 21st century, entrepreneurship taps into these technological advances, opening doors to solutions that did not exist before, driving economic growth and addressing emerging social challenges. Technopreneurship is a result of the strong amalgamation of technology and entrepreneurship, where business visionaries exploit cutting-edge technologies to create new products, Services, and Business models which can scale rapidly and make a significant impact. That is the very essence of technopreneurship, which SpaceX did in the aerospace sector or Airbnb changed hospitality space, and Peloton revolutionised fitness technology. On the rapidly changing landscape of the 21st century – where Industry 4.0, globalisation and digital transformation intersect – governments are needed more than ever to stimulate technopreneurship.

According to the 2023 Global Startup Ecosystem Report, the estimated value of global startup ecosystem is $7. Since startups keep on driving development, employment and financial growth everywhere throughout the world in light of current circumstances, this valuation represents what an important commitment technopreneurship has to the economy as a whole. This creative destruction and this broader economic growth are driven by cutting-edge technology as well as an entrepreneurial spirit. Across the globe, governments will have to evolve their policies and support mechanisms to make it an opportune environment for technopreneurs.

## 8.2 PHILOSOPHICAL PERSPECTIVE

Governments can encourage innovation and new enterprise formation, leading to economic growth and social development by streamlining the processes for registering business operations and offering fiscal incentives. Rawls' theory of justice focuses on making sure that social structures are as beneficial for the most disadvantaged person. In terms of institutions, Rawls posits that a society is fair when it exists in such a way as to benefit the least well-off among its constituents. We see this in initiatives like


DOI: 10.1201/9781003537335-8

digital inclusion, which provides opportunities for other stakeholders who are not at the centre of innovation. Helping to make technology and education more available in underserved regions can improve inclusivity, unlock hidden human capital and support social progress. By guaranteeing that everyone has the chance to engage in the digital economy, governments promote a more equitable and just society.

The Aristotelian notion of the "Golden Mean" posits that virtue lies in the balance between extremes. His philosophy warrants the government to find a balance between the inexorable march of science and the need for regulatory oversight. Excessive government intervention can hamper and limit creativity. However, lax government regulation may pose challenges like data breaches and market failure. This well-rounded approach is crucial for stimulating sustainable innovation, advancing social fairness, and upholding societal cohesion. Triple Helix Model stipulates that active collaboration between academia, industry, and government will act as a driver of innovation and create synergies among sectors (Henry Etzkowitz and Loet Leydesdorff, 1998)

## 8.3 IMPORTANCE OF GOVERNMENT SUPPORT AND POLICIES

Creating an environment that is conducive to new ideas is vital in order to get the government's support, and this has become one of the white-hot issues when it comes to fostering economic growth through technopreneurship. One means to accomplish their goals is the active promotion of technology-based firm creation and growth through legislation. This includes financial incentives, which are also present in the grant packages used by other plans, such as grants for entrepreneurs and subsidies or tax cuts. These policies are common because they allow young firms to focus on innovation rather than capital concerns. Some of the positive steps in this regard have shown how selective government interventions can help promote innovative and entrepreneurial initiatives in technology, with large economic payoffs internationally as well as nationally

Nevertheless, filling the gaps in expertise, funding, and infrastructure is part of supporting technopreneurship by the government. But the truth is that this process can be quite challenging for most startups trying to raise any kind of capital outside their own local region or connect with the much-needed capabilities they need in order to build and scale operations within the current infrastructure. Such disparities will be ameliorated via government policies that furnish financial assistance in a targeted way, improve infrastructure up to world standards and encourage educational/training programs. A successful initiative goes beyond providing the essential resources but establishes an ecology that fosters innovation and sustains over time. Governments are able to address these gaps and accelerate innovation in key areas so that technopreneurs can better improve the economy, as well as governmental technological advancement.

## 8.4 DIFFERENT ASPECTS OF GOVERNMENT SUPPORT

All work on high-tech companies should now include public-private solutions that enable these new forms of enterprise to contribute to economic development, drive sustainability and operate within the complex global digital world. There shouldn't be any surprise that governments are the single most significant driving force of advancement and innovation through setting up hubs for technology, improving young minds in STEM education, or just cutting through all this red tape. Such an approach shows a dedication to growing prosperity inclusively yet at the same time preserving social cohesion. In the age of swift technological strides, states capacitate and equip innovators to set up the best vision for tomorrow. Governments around the world have employed various ways of providing tech business support – at different scales depending on levels, capabilities and needs. Components of government support. Some of the aspects this support are as follows.

### 8.4.1 Financial Support

One of the most critical factors in promoting innovation, growth and greater technopreneurship is government support. This will reduce many barriers startups have, as most of the governments are financially assisting them in a number of ways premiere enabling its citizens to work on innovation with less worrying about profits. The financial support mechanisms, such as grants, subsidies and tax incentives are provided to reduce the cost of developing new technologies in order to ease enhancing businesses investment into R&D (research and development). Thus, these financial instruments not only allow startups to survive the often-tumultuous early stages but also foster a culture of innovation and collaboration and provide room for growth. Government support for technopreneurship can thus trigger economic growth, advance national ranks in the global market arena and achieve widespread diffusion of technological benefits.

### 8.4.2 Grants and Subsidies

It has been recognized by governments all over the world that technopreneurs need to be supported with finance if we want cutting-edge innovation and sustainable growth. Historically, venture capitalists and private equity firms are the main private players in supporting such efforts. But with a recognition of the strategic importance of fostering innovation, governments have increasingly taken a more proactive role. The active involvement of the government manifests the sheer need and urgency. The grants and subsidies through government exchequer not just place more money directly into the pockets of at least some startups and freeing them from immediate pressure to turn a profit, allowing the focusing of resources on R&D but also creates a level playing field where the startups can cushion themselves against the mighty technopreneurs in the market. For example, the United States' Small Business Innovation Research program provided competitive grants to small businesses for research and development projects that were recognized as having great commercial potential. Similarly, the European Union's Horizon 2020 program provides funding for an extremely broad range of R&D activities, from scientific research to the development of frontier technologies, in a budget of close to €80 billion over seven years.

Such initiatives are supported by research through various studies. For instance, Wallsten (2000), evaluated the SBIR program. He found that such grants significantly

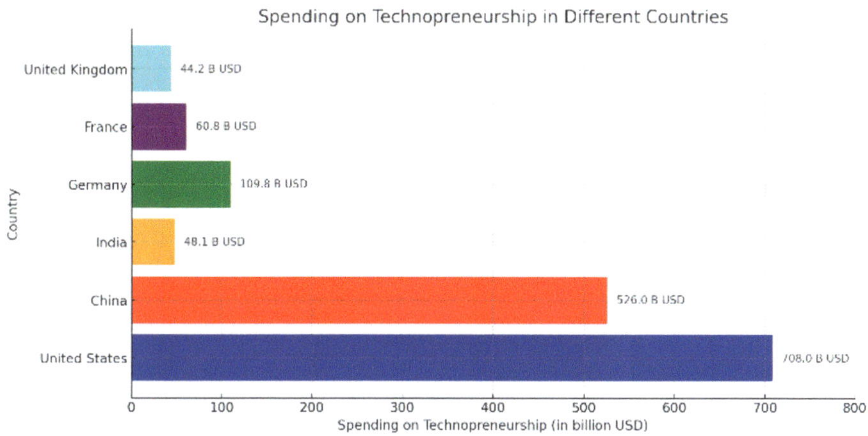


**FIGURE 8.1** Estimated Global Startup Ecosystem Value (2023). (Global Startup Ecosystem Report, 2023.)

raised the chances of small companies being involved in innovative activities and hence concluded that government funding can complement private investment effectively. The paper by Cumming and Fischer (2012) considered the effect of R&D grants on the innovation output of small firms and found evidence that directly showed government subsidies are crucial in filling the finance gap in early-stage technological startups. Takalo, Tanayama, and Toivanen (2013) also find evidence that such subsidies end up resulting in strong increases in the quantity and quality of innovative projects pursued by firms in their study of the Finnish R&D grant system. Figure 8.1 illustrates the estimated global start-up ecosystem value, emphasizing the economic contributions of technopreneurship. It shows that there is a stark difference between the spending patten observed in developed countries as well as higher-middle income countries like China as compared with India. China and the USA spend almost eleven and fifteen times more money on technopreneurship as compared with India at a macro-level. The per-capita technopreneurship expenditure of India is considerably low as compared with developed countries in the world.

### 8.4.3 Favourable Tax Structure

The primary purpose of tax incentives is to lure domestic and foreign investment in the country. Many countries of the world result in giving tax breaks with the aim of reducing the cost of production of the domestic production and ultimately give a competitive edge against foreign counterparts. The benefits are offered through various ways such as tax breaks, deductions, and credits. The governments build more conducive financial environments for their technopreneurs so that they can continue investing in the development and research of new technologies. For instance, Singapore has massive tax deductions related to R&D expenses as part of its tax incentive scheme, making it more lucrative to invest in innovation for businesses. Tax deductions reduce the income subject to taxation by allowing companies to subtract a portion of their R&D expenses from their taxable income. Ireland, known for its competitive corporate tax rates, also provides additional R&D tax credits. Unlike deductions, tax credits directly reduce the tax owed, offering a dollar-for-dollar reduction in tax liability. This has successfully attracted numerous tech companies to establish operations and fostered a vibrant technopreneurial ecosystem.

Research demonstrates the efficacy of these tax incentives. A study conducted by Bloom, Griffith, and Van Reenen (2002) emphasizes that R&D tax credits have a significant positive impact on R&D spending, resulting in a higher level of innovation output. Another paper authored by Hall and Van Reenen (2000) examines the influence of fiscal incentives for R&D and concludes that such policies generally succeed in stimulating greater private sector investment in innovation. Furthermore, Lokshin and Mohnen (2012) present evidence from the Netherlands, indicating that R&D tax incentives lead to a substantial rise in the amount of private R&D investment.

There has been a plethora of policies and programmes implemented in major economies of the world to promote technopreneurship through tax incentives. Indian government initiated the start-up India scheme, which offers a three-year tax holiday for startups along with exemption from paying capital gains tax. Relaxation in their compliance with environmental and labour laws has also reduced their regulatory burden. High and New Technology Enterprise (HNTE) Program implemented by the Chinese government in 2008 offered a substantial reduction in the tax liability for companies investing in high-tech innovation and improving the R&D in the country. The firms investing in high-tech firms will only attract a 15% tax as compared with the standard tax rate of 25%. The program has been a crucial part of China's strategy to transition from a manufacturing-based economy to an innovation-driven one (KPMG, 2021). The research and development (R&D) Tax Credit program in the US was initiated in 1981. It was designed to give tax credit for qualified R&D investments. It is clearly discernible that the developed world started tax incentives to boost innovation in their economies at an earlier stage as compared to developing countries.

It is to be noted that tax incentives should create incentives for innovation and spur real infrastructure in the economy. However, an inefficiently designed tax structure may favour mighty businesses at the cost of small businesses. The robust and efficient tax framework is required to avoid rent-seeking, as noted by eminent economist Joseph Stiglitz in his book Freefall: America, Free Markets, and the Sinking of the World Economy.

### 8.4.4 Infrastructure Support and Development

Investments in infrastructure often have significant positive spillover effects on other sectors of the economy. Improved transportation infrastructure, for example, can reduce costs for businesses, enhance productivity, and open up new markets (Aschauer, 1989). Additionally, such investments can lead to increased economic activities and innovation by providing the necessary support systems for businesses to thrive (Gramlich, 1994).

In this regard, various governments around the world have felt the growing need to develop infrastructures that support technopreneurship. Tech parks and innovation hubs play a very relevant role in this respect by providing both physical and intellectual resources which facilitate innovation. For instance, Silicon Valley in the United States is probably one of the best exemplary tech ecosystems anywhere in the world, with the added advantage of strong infrastructure and networking opportunities that have successfully reared several startups and established firms alike. This can partly be attributed to various funding initiatives from the federal government into R&D, such as the National Science Foundation and grants from DARPA, and to provide fiscal wherewithal necessary for stimulating innovation and technological development. As Elon Musk, the CEO of Tesla and SpaceX, says, "Silicon Valley has a very powerful network effect; it's a very positive feedback loop," in order to emphasize how big a role supportive infrastructure plays in fostering innovation. Zhongguancun in China, often referred to as China's Silicon Valley, is home to a large number of tech startups and established companies. It provides a dynamic environment that fosters innovation. The Chinese government has implemented the "Innovation-Driven Development Strategy," which involves significant investments in research and development (R&D) and innovation parks. The Zhongguancun Science Park has received substantial funding and policy support to promote technological innovation and entrepreneurship. This demonstrates how targeted government intervention can create a thriving tech ecosystem. Jack Ma, the founder of Alibaba, has noted that innovation in China is now more encouraged than ever before, and it is reshaping the country's economic landscape. This highlights the transformative impact of government support (Feldman & Florida, 1994).

According to the World Bank, "digital infrastructure helps in creating a knowledge economy by providing the necessary tools and platforms for innovation, information exchange, and efficient business processes" (World Bank, 2016). The Digital India initiative by the Indian government laid the foundation stone of digital infrastructure in the country, which involves the establishment of Broadband Highways, online delivery of public services and promotion of digital literacy. Atal Innovation Mission (AIM) is designed to create an environment of innovation through the creation of incubation centres and tinkering labs. International Tech Park in Bangalore and the T-Hub in Hyderabad manifest the outcome of the efforts by the successive state and federal governments in promoting technopreneurship especially among the startup ecosystem in India. Policies like the Software Technology Parks of India (STPI) scheme and the establishment of Special Economic Zones (SEZs) have been instrumental in increasing exports, promoting investments and expanding the infrastructure base of India. These parks receive fiscal incentives, including tax exemptions and infrastructure development support, which provide state-of-the-art facilities and a supportive environment for tech entrepreneurs.

Contemporary enterprises that are founded on technology are dependent not just on physical infrastructure but also on digital infrastructure. Countries that make substantial investments in more sophisticated digital infrastructure generate an atmosphere that is more conducive to the growth of technological entrepreneurs. Take South Korea's 5G network as an example; it provides the kind of lightning-fast internet that cutting-edge IT companies need to unleash their full potential. The South Korean government has made major efforts in building this infrastructure, including programs such as the "K-ICT Strategy 2016," which dedicated significant cash to improve the country's digital capabilities and create a robust and future-proof infrastructure. Creative Economy initiative by South Korean government is also a notable example. Creative Economy Innovation Centers, public-private partnerships, financial incentives, and support for digital startups and innovation are all parts of South Korea's Creative Economy Initiative, which aims to encourage technopreneurship.

In his prediction that the greatest digital infrastructure will be the driving force behind the next wave of innovation, Eric Schmidt, the former CEO of Google, brought to light the crucial relevance of digital infrastructure in the process of promoting creativity among individuals.

### 8.4.5 Human Resource Support and Development

The HR support established the groundwork for innovation by equipping individuals with the necessary knowledge, skills, and mindset to excel in the rapidly changing technological landscape. Figure 8.2 illustrates the steady increase in total global spending R&D from 1996 to 2022. This upward trend highlights the growing emphasis on innovation and technological advancement across the world, with spending reaching over 2,475 billion PPP (Purchasing Power Parity) U.S. dollars in 2022.

### 8.4.6 Educational and Skill Development Programs

On global scale nations acknowledge investing in education. Specifically in Science, Technology, Engineering and Math or the STEM fields. This is essential.

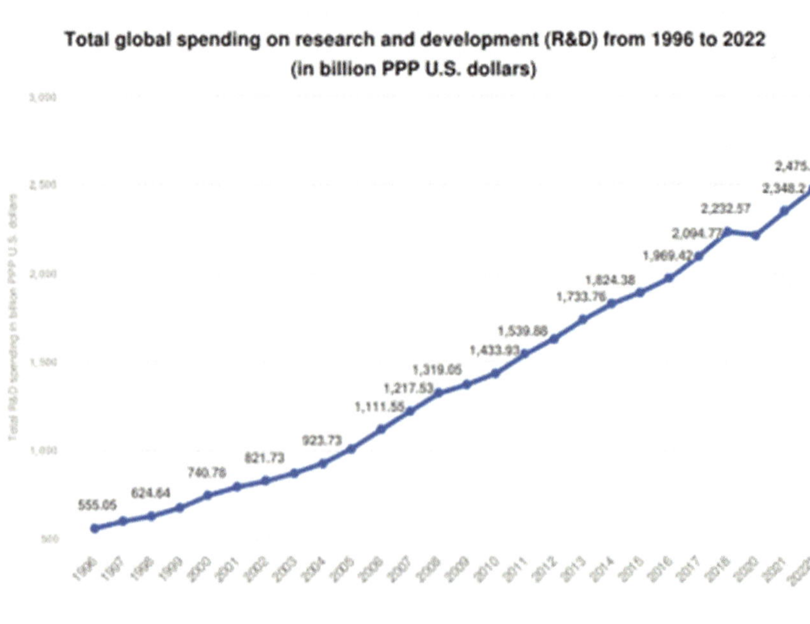


**FIGURE 8.2** Total Global Spending on Research and Development (R&D) from 1996 to 2022. (UNESCO Institute for Statistics; R&D World; Statista 2024.)

Along with this, comprehensive entrepreneurship training is important. This will nurture next generation innovators and entrepreneurs. These initiatives make sense. They cultivate a highly skilled workforce. They drive economic growth. Additionally, nations can enhance competitive advantage. Widespread benefits happen from technological advancements. Important research underlines the value of these programs.

### 8.4.7 STEM Education

Governments globally are pouring resources into STEM education. They are doing this to spur innovation. Powerful STEM education equips students with essential critical thinking skills. It also provides them problem-solving skills. These skills are needed to outperform in today's rapidly changing technological world. A unique system in Germany integrates vocational training with academic learning. It is a dual education system. It creates powerful pipeline of skilled workers. These workers are primed for the rigor of tech industry. The system melds practical experiences with theoretical knowledge.

Upon graduation, students from this system are ready for industry. The German government lends robust backing to this dual system of education. They make substantial investments yearly. These billions go into vocational education. They also fund training programs. Angela Merkel had this to say about the German educational process: "Engineering and manufacturing in Germany are built on sturdy education and training." In a highly successful Finnish curriculum, STEM is central. The focus here is creativity and problem-solving. Inspiring creativity is crucial to everyone here especially the students. Also directly targeted is the student's ability to resolve practical problems. Education in Finland is renowned for its focus on individual students. The goal is to instil a deep-rooted curiosity for learning and exploratory behaviour. Interesting is the fact that Finnish government allocated a significant investment. The sum goes up to 5.5% of the country's gross domestic product. This big money goes into the education sector. The objective is high-quality educational opportunities for every student. A Finnish education reformer named Saku Tuominen insists on something. The belief is that preserving creativity and curiosity is pivotal. This enables the fostering of innovation. These innovative educational methods wield much influence. Therefore, students learn more than just knowledge. Equally important is their enhanced skill application. It's developed in novel and hands-on ways.

### 8.4.8 Entrepreneurship Training

Entrepreneurship training programs are a prerequisite to unleash the true potential in any economy. They equip individuals with essential skills for business creation and its expansion. These are additions to STEM education. Over a global stage, governments attempt to run many programmes to skill the available demographic dividend in their respective countries. In Israel, the Yozma program exists which offers training and guidance to the youth. Yozma helps in navigating business establishment intricacies. And it has indeed shown results for Israel.

Israel became a nation famous for a burgeoning startup ecosystem through a culture of innovation and risk-taking. Substantial financial support goes to Yozma. That comes from the Israeli government. It first provided one hundred million dollars of funding. Consequently, it attracted private investment.

United Kingdom put in place several measures to back up entrepreneurship. One of them is the Start Up Britain campaign. This initiative provides valuable resources and help new businesses. These come in the form of mentorship, funding opportunities and networking events. With regards to finance, UK government offers comprehensive support. This includes many things like grants and tax incentives. One big player is the £2.5 billion British Patient Capital program. It's a dedicated fund meant to increase innovation growth in the country. This cultivates innovation and economic growth. Take for instance Richard Branson, founder of Virgin Group. He holds an important position in supporting budding entrepreneurs. This kind of backing enables them to triumph over obstacles. And to achieve notable success.

Hanushek and Woessmann (2015) embarked on a study. They found that countries with strong STEM education systems generally see high levels of innovation and economic growth. In another standpoint, discovered by Fairlie and Robb (2007) entrepreneurship training plays a vital role. It enhances business outcomes and supports economic development. The findings reflect the fundamental role of education and training. They help in building a competitive and innovative economy. Regulatory Framework refers to set of rules. These regulate a specific industry or sector. It provides structured setup. This ensures adherence to standards regulations within industry form a robust setup. Regulatory setup fits well with technopreneurship. It also provides legal and institutional basis. This aids business activities. It also protects intellectual property and importantly ensures fair competition. Governments sit at the centre of this regulatory framework, and it facilitates a transparent regulatory environment.

Regulatory Framework is understood as rules and regulations that exist around an industry or sector constitute what could be referred to as the regulatory framework. It gives a structured way of ensuring compliance and standards within such an industry. A strong regulatory framework breeds an enabling environment for technopreneurship to welcome upscale in growth. Effective regulation provides appropriate legal and institutional foundations, which assure business venture activities, protection of intellectual property, and fair competition. These include developing a clear and efficient regulatory environment-one that can reduce barriers to entry, encourage innovation, and thereby attract investment in its many forms, both domestic and international. Theories of regulation, such as the Public Interest Theory and the Capture Theory, provide further detail on the purposes and impacts of regulatory policies. The Public Interest Theory posits that regulations are supposed to safeguard people's welfare and ensure that economic activities are conducted in a manner that is just and fair. Conversely, the Capture Theory countered that regulations may veer too much toward heavy influence from industry interests. A balanced approach between both perspectives is valuable in formulating policies that would, in fact, be for the benefit of the general economy and society as a whole.

## 8.5 DIFFERENT ASPECTS OF 'POLICY FRAMEWORKS'

### 8.5.1 Ease of Doing Business (EODB)

Capture Theory and the Public Interest Theory propose effectiveness of the legislative system. According to Public Interest Theory, rules level the market and protect the general people. In Denmark and New Zealand, rules are clear-cut. According to the Capture Theory, the sector under control could influence the legislation. Transparency and efficiency are underlined to prevent too strong influence and guarantee that laws help the economy.

Technopreneurship is best suited in a regulatory setting free of bureaucratic obstacles. Simplified rules reduce company administrative time so they may grow and be creative. For example, New Zealand is well-known globally for its quick and effective corporate registration system. Online registration and simplified compliance requirements have made it easier for New Zealand entrepreneurs to start and run firms. Thus, New Zealand routinely ranks high in the World Bank's Ease of Doing Business Index. Former New Zealand Prime Minister John Key stressed the government's goal of making it easier for businesses to start and grow. Denmark has an effective business regulatory structure. The Danish government has simplified tax filing, reduced administrative procedures, and clarified regulations, which have considerably reduced Danish company expenses and complexity. Former Danish Minister for Economic and Business Affairs Morten Ostergaard says Denmark's policies promote growth and innovation. Danish technopreneurs benefit from legislation that assists their core business operations and growth.

The World Bank's Ease of Doing Business (EODB) Index assesses regulatory efficiency and business friendliness for various nations. This index emphasises streamlined regulations. High-ranking countries like New Zealand and Denmark show how simplifying company procedures and reducing administrative barriers can boost entrepreneurial activity and economic growth (Djankov et al., 2002).

### 8.5.2 Intellectual Property Rights (IPR)

Protecting IP (Intellectual Property) is essential and indispensable for the honest growth of technopreneurship policy. Strong intellectual property rules let inventors protect and profit from their ideas, increasing R&D investment as a result. The US protects intellectual property with strong patent, trademark, and copyright laws. Technopreneurs' innovations are protected by the USPTO. This robust intellectual property framework spurs innovation and technology investment. Thomas Edison said an idea's worth is in its implementation. Through the European Patent Office (EPO) and Community Trademark system, the EU protects IP. These institutions defend intellectual property between member states, making it easier for technopreneurs to secure their discoveries in Europe. Researchers and developers can invest in their work knowing that the EU will defend their IP. Former European Commission President Jean-Claude Juncker stressed the importance of intellectual property rights in promoting innovation and economic growth in Europe.

Research shows that well-designed regulatory frameworks can boost business growth and inventiveness and concludes that simpler business laws increase entrepreneurship and economic activity. They found that strong intellectual property rights encourage innovation and R&D investment. The role of regulatory frameworks in supporting technopreneurs is critical, decisive, and seminal.

## 8.6 RECENT GLOBAL TRENDS IN TECHNOPRENEURSHIP SUPPORT

The developed world prioritises high-tech startup support. Advanced digital infrastructure, robust IP protection, and large financial incentives do this. Advanced nations like the US, Germany, and Japan invest heavily in digital infrastructure and innovation centres. For instance, the US National Network for Manufacturing Innovation (NNMI) promotes innovative manufacturing technology through government-private sector partnerships. This program provides a common space for R&D and product unveils to hasten and promote innovation. Technopreneurs must be able to protect their creations and ideas. IP laws must be impeccable to safeguard innovations legally. For instance, the EU has implemented the Digital Europe Programme (DIGITAL). Supercomputing, AI, cybersecurity, and advanced digital skills are prioritised in the Digital Europe Programme. To encourage new technology investments, digital capabilities and IP protection are improved as shown in Table 8.1.

As shown in Table 8.1, developed nations prioritize advanced infrastructure and large-scale financial incentives, while developing countries focus on foundational initiatives like digital literacy and basic connectivity. For technopreneurship to thrive, these factors are essential. Digital India is a campaign launched by the Government of India to make its services available to citizens electronically via improved online infrastructure and by increasing internet connectivity. This effort is crucial to eliminating the technology access gap and helping more people to participate in the digital economy. Locally designed programs like microfinance for small company owners and community-led innovation projects are crucial. Mobile banking service M-Pesa has transformed financial inclusion in Kenya. It has enabled mobile payments and microloans to help people establish and grow businesses in regions without traditional banking facilities. Technology entrepreneurs have benefited from M-Pesa's support.

**TABLE 8.1**
**Global Trends in Technopreneurship Support – A Comparison**

| Developed Nations | Developing Nations |
|---|---|
| • High Tech Startup Support<br>• Advanced Digital Infrastructure<br>• Robust IP Protection<br>• Large Financial Incentives<br>• US: National Network for Manufacturing Innovation (NNMI); EU: Digital Europe Programme. | • Building Foundational Infrastructure<br>• Improving Digital Literacy<br>• Enhancing Internet Connectivity<br>• India: Digital India Campaign; Kenya: M-PESA policy. |

Despite attempts, many issues remain. These hurdles include regulatory issues, financing constraints, skill shortfalls, and worldwide rivalry. Excessive rules can inhibit innovation. Therefore, a balance between control and flexibility is needed to protect consumers and improve technology. Funding is difficult, especially in poor nations with minimal venture capital. Local investment ecosystems and government-backed loans or guarantees are needed to fix this. Globally, key technology fields lack qualified labour. Investment in education and training is needed to enable innovation and economic growth. Technology entrepreneurs compete fiercely in a globalising environment. Thus, helping small and medium-sized firms grow and compete internationally is crucial. Creating a thriving and competitive market requires this assistance. Economic growth theories, like the Innovation Systems Theory, emphasise innovation-friendly environments. This idea highlights the role of governments, corporations, and educational institutions in promoting technological innovation and economic prosperity. In a 1992 study on national innovation systems, Lundvall stressed the importance of government policies in fostering innovation.

Research shows that these supporting strategies have great impact. In 2012, Acemoglu and Robinson underline the need of inclusive institutions for both creativity and economic progress. These results underline the need of thorough support systems for technological entrepreneurship. By tackling these issues and using support systems, governments can help to build a more dynamic and inclusive environment for technological entrepreneurs. Technical progress, national competitiveness, and fair innovation distribution depend on this approach. *"Innovation is what sets leaders apart from followers."*—Steve Jobs.

High-tech businesses need a friendly environment to support creativity and keep their competitive edge in the worldwide market. The United States, Germany, and Japan admit this and fund greatly in research institutions and digital infrastructure. By means of government-private sector partnerships, the US National Network for Manufacturing Innovation (NNMI) promotes advanced manufacturing technologies. To speed innovation, this project provides a common environment for research and development as well as commercialization. Technopreneurs have to protect their ideas and creations. Intellectual property laws have to be strong to defend inventions in a moral and legal sense. Illustration: European Union's Digital Europe Program. Supercomputing, artificial intelligence, cybersecurity, and advanced digital skills top this curriculum. It aims to improve intellectual property protection and digital skills to boost new technology investment.

However, developing nations prioritise foundation building. This includes improving digital infrastructure and education. These places emphasise internet connectivity and digital literacy, which are essential for technopreneurship. Initiatives in the same direction improve online infrastructure and internet connectivity to make government services more accessible to citizens. This effort is crucial to eliminating the technology access gap and helping more people to participate in the digital economy. Locally designed programs like microfinance for small company owners and community-led innovation projects are crucial. It also transformed financial inclusion in developing countries. This mobile banking service has enabled mobile payments and microloans, enabling people to start and grow their companies in areas without conventional banking facilities.

Despite attempts, many issues remain. These problems include regulatory issues, financing issues, skill shortfalls, and global rivalry. Excessive rules might hinder innovation. To protect customers and advance technology, regulation and freedom must be balanced. Overly severe data protection rules may slow AI technology. To protect consumers and encourage innovation, regulation and flexibility can be balanced. Funding is difficult, especially in poor nations with minimal venture capital. Many African businesses struggle to attract foreign investors due to perceived dangers. Fostering local investment ecosystems and government-backed loans or guarantees can help solve this challenge. Additionally, critical technology skills are scarce worldwide. Digital firms may be hampered by the worldwide cybersecurity skills gap. Investment in education and training is crucial for innovation and economic success. Technology entrepreneurs face severe competition in a globalising market. Small startups often struggle to compete with digital giants. To promote a lively and competitive market, SMEs (small and medium-sized enterprises) must be helped to expand and compete globally.

The theories of economic development, such the Innovation Systems Theory, highlight the need of encouraging creativity. This approach stresses how governments, businesses, and educational institutions could help to improve technical and financial growth. An analysis of national innovation systems published in 1992 underlined the need of government initiatives in fostering creativity. These kind of helping deeds have paid off. In their 2012 paper, Acemoglu and Robinson highlight the need of inclusive institutions for economic development and creativity. These results highlight the need of wide support systems for technopreneurship. Governments might offer a more dynamic setting where technopreneurs might negotiate obstacles and make use of support systems. Modern technology, national competitiveness, and fair sharing of invention depend on this approach.

## 8.7 NEED FOR GLOBAL CONSENSUS ON TECHNOPRENEURSHIP

Dealing with the difficulties ahead calls both international agreement and cooperation rather highly. International agreements on digital commerce, internet security, and intellectual property protection might give technopreneurs all around more possibilities and tools. The restrictions of digital commerce draw attention to the worldwide struggle to solve significant problems resulting from lack of consensus. Established by the European Union (EU), the General Data Protection Regulation (GDPR) lays a strict benchmark for data privacy applicable to businesses all over. For companies functioning in both the US and the EU, data privacy laws have presented difficulties. Further aggravating these issues in 2020, the European Court of Justice withdrew the Privacy Shield agreement.

This choice caused major interruptions for several businesses depending on data flows between the US and the EU. Emphasizing the importance of current worldwide norms and agreements to facilitate digital trade is crucial. These agreements could help to solve these problems, therefore simplifying international trade and supporting the world economy.

Disparities in the rules and laws that control the transfer of data and online transactions can be a problem for multinational corporations. Companies aiming to expand internationally face formidable obstacles due to the widely differing data privacy and security regulations between nations. One possible solution to these problems is the creation of global agreements and standards to ease digital trade. Simplifying cross-border transactions and promoting greater global economic integration can be achieved through the adoption of shared frameworks among nations.

Globally, cybersecurity is a very important issue that calls for cooperation to properly address. The cooperative efforts of governments addressing cybersecurity issues are best shown by the Budapest Convention on Cybercrime. Dealing with online and computer crimes, this pact is a big success. This project aims to promote worldwide cooperation by means of harmonizing legal systems, improving investigative methods, and so supporting nation-wide cooperation. Not all nations have committed to this convention; hence it is crucial to underline the need of more worldwide cooperation to create and implement effective cybersecurity policies. Working jointly, several parties can design thorough cybersecurity systems. These systems can help to lower vulnerabilities and improve the general security of technopreneurs' enterprises.

Furthermore, IP rules vary between nations, which can provide challenges for multinational companies. Establishing basic rules for several forms of intellectual property control for people from nations members of the World Trade Organisation (WTO), the TRIPS agreement is a thorough international agreement. TRIPS notwithstanding, there are still notable differences in the application of intellectual property rights throughout nations. One may make an interesting analogy between American patent laws and Chinese ones. These variations have caused strife and presented difficulties for worldwide corporate and innovation. Global consistency in intellectual property laws and enforcement systems is absolutely vital if we are to protect ideas and advance technological entrepreneurship. Consistent policies to safeguard intellectual property help nations establish a safer environment for businesses and innovators. This therefore promotes intellectual and worldwide economic progress as well as fresh ideas development.

Establishing a global accord and fostering cooperation is essential in tackling these challenges and cultivating a conducive atmosphere for technopreneurs. International agreements and standards across diverse domains, including digital trade, cybersecurity, and intellectual property rights, are crucial for establishing a stable and inclusive environment that fosters innovation. This will assist entrepreneurs in achieving success on a global scale.

## 8.8 WAY FORWARD FOR THE GOVERNMENT

Technopreneurship requires flexibility and adaptability. Technology advances quickly, therefore government policy must be adaptable. Stable policies limit innovation and progress. Thus, governments must design flexible systems that can adapt to new technology and market changes. Keeping up with global technological advances and allowing technopreneurs and other specialists to influence policymaking is crucial. Governments can create adaptive regulatory sandboxes to better understand and apply legislation. These sandboxes allow real-world testing of new technologies without risk. Flexible frameworks that enable current and innovative policies can help governments support technopreneurship in a changing environment.

Teaming up to improve technopreneurship. Fostering public-private collaboration and international cooperation is vital. Governments need private sector help to overcome modern technology's complex difficulties. People may boost economies and technology by collaborating on research, funding, and innovation centres. The German government, industry leaders, and educational institutions formed Industry 4.0 to foster manufacturing innovation. International relationships that promote research and development can provide technology entrepreneurs with a worldwide edge. Horizon Europe is an EU program. Due to information and resource sharing, these relationships boost the technopreneur ecosystem and spur innovation.

Emphasizing universal access and inclusivity and fighting for inclusivity through policies is crucial to ensuring equal access to technopreneurship, especially for excluded groups. Governments must establish and implement technology industry diversity and inclusion rules. Educational programmes to teach poor communities digital skills, investment programmes for women-led enterprises, and minority technology industry engagement are examples of initiatives. The US Minority Business Development Agency supports minority-owned technology companies. For innovation and economic success, governments should foster inclusion, which will create a diverse and vibrant technopreneurial ecosystem.

Lastly, embracing sustainability makes sure that progress in technology is good for both people and the earth. Policies for technopreneurship should put ecology first. The Green Transition Fund gives money to green technology companies to help Denmark reach its goal of being carbon neutral by 2050. To encourage technopreneurs to make inventions that are good for people and the environment, government aid programs should include sustainability concepts. We can reach this goal with the help of projects that build a circular economy, strict environmental rules, and incentives for green technology. These rules also encourage smart new ideas that help the economy grow without hurting people or the environment. If states put sustainability first, eco-entrepreneurship and new technology can work together.

## 8.9 CONCLUSION

Government intervention in the form of law and policymaking for technology entrepreneurs is intertwined with many other parallel factors, as it is complex and dynamic. In general, cooperative and friendlier structure and access to monetary schemes, infrastructure and facilities, and transparent and efficient regulatory environment are seminal for any technopreneur to succeed. In addition to that, Innovation, Economic Growth, & Healthy Development of all, including LDCs (Least Developed Countries), is a challenge in front of governments around the world. The world is like a global village in the age of AI; cooperation & collaboration on future ideas with the government will be a panacea. Government of the future need to be fast and agile in response to changing technologies and commerce. World Economic Forum's founder Kalus Schwab once remarked, "Technology will change our Life, Work, and Social connections". It offers us the clarity that more than Hard and Soft power, future governments will require Fast Power, which is agility and adaptability in policy making.

Similarly, Digital trade, Cybersecurity, and IPR (intellectual property rights), and Consumer Protection cannot be ignored. Government of the future must find a way to prioritise the issues with changing technology at a global scale. Global agreements have the potential to alleviate and support technopreneurs. But merely signing a global convention might not work; sincere follow-up and monitoring are expected.

Government must proactively train the teams for upcoming challenges. For this purpose, education & training must be imparted in curriculum and formal education to bridge the skill gap. Though the steps sound long-term and idealist, it is the starting point for the preparation of the future workforce in the digital era. Industry demand and contemporary skills should be considered while formulating curricula in the educational institutes. Moreover, ensuring equal access to digital resources for women, minorities, and subaltern classes will foster innovation and promote societal advancement. Governments must ensure the equitable distribution of technological advantages. Development should go hand in hand with social equity. This is not for any benevolence but to main fairness in the approach.

In the book *The Politics*, Aristotle, the pioneering philosopher who guided how to run a government, discussed the idea of the "Golden Mean". This idea emphasises a moderate and balanced approach rather than working with extremes. Extremes in the Government never work; we need ideas that are practical in nature and accepted by society. Golden Mean is sustainable in nature as it considers the interests of all; hence, it is the best approach to formulate any policy. It helps in continuing the policy forward in changing and growing global economy where imagination drives long-lasting progress and improves life in general. Governments, technopreneurs, and international organisation in the comity of nations working together as a one single team will help to solve digital issues and forward technology.

## REFERENCES


Aschauer, D. A. (1989). Is public expenditure productive? *Journal of Monetary Economics*, 23(2), 177–200.

Bloom, N., Griffith, R., & Van Reenen, J. (2002). Do R&D tax credits work? Evidence from a panel of countries 1979–1997. *Journal of Public Economics*, 85(1), 1–31.

Bresnahan, T., Gambardella, A., & Saxenian, A. (2001). Old economy inputs for new economy outcomes: Cluster formation in the new Silicon Valleys. *Industrial and Corporate Change*, 10(4), 835–860.

Cumming, D., & Fischer, E. (2012). Publicly funded business advisory services and entrepreneurial outcomes. *Research Policy*, 41(1), 34–44.

Djankov, S., La Porta, R., Lopez-de-Silanes, F., & Shleifer, A. (2002). The regulation of entry. *The Quarterly Journal of Economics*, 117(1), 1–37.

Fairlie, R. W., & Robb, A. M. (2007). Families, human capital, and small business: Evidence from the characteristics of business owners survey. *Industrial and Labor Relations Review*, 60(2), 225–245.

Feldman, M., & Florida, R. (1994). The geographic sources of innovation: Technological infrastructure and product innovation in the United States. *Annals of the Association of American Geographers*, 84(2), 210–229.

Etzkowitz, H., & Leydesdorff, L. (1998). The triple helix as a model for innovation studies. *Science and Public Policy*, 25(3), 195–203.

Ginarte, J. C., & Park, W. G. (1997). Determinants of patent rights: A cross-national study. *Research Policy*, 26(3), 283–301.

Gramlich, E. M. (1994). Infrastructure investment: A review essay. *Journal of Economic Literature*, 32(3), 1176–1196.

Hall, B. H., & Van Reenen, J. (2000). How effective are fiscal incentives for R&D? A review of the evidence. *Research Policy*, 29(4–5), 449–469.

Hanushek, E. A., & Woessmann, L. (2015). *The knowledge capital of nations: Education and the economics of growth.* MIT Press.

KPMG. (2021). *R&D tax incentives: Global comparison report.* KPMG International.

Lokshin, B., & Mohnen, P. (2012). How effective are level-based R&D tax credits? Evidence from the Netherlands. *Applied Economics*, 44(12), 1527–1538.

Lundvall, B. A. (1992). *National systems of innovation: Towards a theory of innovation and interactive learning*. Pinter Publishers.

Takalo, T., Tanayama, T., & Toivanen, O. (2013). Estimating the benefits of targeted R&D subsidies. *Review of Economics and Statistics*, 95(1), 255–272.

Wallsten, S. J. (2000). The effects of government-industry R&D programs on private R&D: The case of the Small Business Innovation Research program. *The RAND Journal of Economics*, 31(1), 82–100.

World Bank. (2016). *World development report 2016: Digital dividends.* Washington, DC: World Bank.

# 9 Technopreneurship in India

## *A Case Study Analysis of Success Factors and Challenges With Reference to Bangalore, Karnataka*

*Anand Shankar Raja M, L. Mynavathi, Smita Kavatekar, Kiran Kumar M, and Akshay Kumar Annam*

## 9.1 INTRODUCTION

In the highly competitive landscape of modern business, many aspiring entrepreneurs fail due to a lack of a unique selling proposition (USP) and the presence of competitors offering superior products and services. However, a distinct category of entrepreneurs, known as technopreneurship, focuses on technical innovation, ensuring their start-ups contribute significantly to economic growth and development through digital technologies. These technopreneurship possess not only business, leadership and creative skills, but also technical expertise, setting them apart from traditional entrepreneurs. Technopreneurs are entrepreneurs who leverage their technical skills to create business opportunities and contribute to the digital economy's development. They possess a deep understanding of market opportunities, government expectations, and realistic consumer needs, particularly in the digital age where technology-aligned products and services have a profound impact. The integration of technology into entrepreneurship differentiates itself from traditional entrepreneurship due to the significant role of technology. Technopreneurs have to strive very hard to maintain and manage Intellectual Property Rights (IPR) which makes their business very unique from the other competitors in the market. Because technopreneurs bring a lot of creativity in their products and services which have to be further protected through various IPR related services such as patents, copyrights, trademarks etc. This process involves a lot of challenges and has many complexities involved. Crossing this hurdle and sustaining in the market as a unique leader is a real time challenge facing the majority of the technopreneurs. However, being innovative and creative will be a

DOI: 10.1201/9781003537335-9

business growth driver for the technopreneurs says (WIPO, 2020). The Government of India actively supports Technopreneurship through various initiatives.

In the Indian context, many technopreneurs are highly benefited due to the diligent support of the government. For instance, the PRISM Program introduced by the Ministry of Science and Technology provides support through funds, training programs, awareness campaigns and paves way for the young aspirants to be "Technopreneurs" in various areas in various industries and sectors. In addition, apart from the PRISM program few other initiatives such as the MUDRA, PMMY and RKVY also offer financial and non-financial support to the technopreneurs (Department of Science and Technology, 2021; PMMY, 2021). These supports help the technopreneurs explore new opportunities, capture the market, contribute towards the socio-economic development of the nation. Though the technopreneurs may pass the initial hurdles in their technopreneurial journey, they still face other issues and challenges with regard to the changing government rules and regulations. Moreover, every day is a new day for the technopreneurs as they have to invest maximum time and budget on technological upgrades, If not, they may fail and decline. Many popular technopreneurs across the globe have also faced similar challenges and these challenges will always exist. These challenges are sometimes the driving force to succeed.

Other notable technopreneurs include Jennifer Doudna with CRISPR gene-editing technology, Nervana Ramnath's affordable diagnostics, David Ishee's bioprinted organs, Demis Hassabis's DeepMind, Amara Raja Chelli's sustainable bioplastics, and Vinisha Umashankar's solar-powered ironing cart.

These innovations illustrate the diverse impact of technopreneurship on solving global challenges. Universities and higher educational institutions (HEIs) play a crucial role in fostering technopreneurship. They serve as hubs for creating environments that support technopreneurs through curriculum and pedagogical changes, multidisciplinary research, incubation centres, and mentoring programs. By promoting interdisciplinary collaborations and maintaining healthy relationships with regional stakeholders, HEIs can significantly contribute to regional and national development through application-based research and development. Governments and policymakers must continue to support technopreneurship to achieve sustainable economic development. By addressing financing, IPR protection, and regulatory challenges, they can create an environment conducive to innovation. Sustainable development challenges can be tackled through technopreneurial solutions, ensuring balanced growth and long-term prosperity.

Indian technopreneurs like Nithin and Nikhil Kamath, co-founders of Zerodha, have revolutionized the financial services sector with a low-cost, transparent brokerage model. Similarly, Sachin and Binny Bansal, founders of Flipkart, transformed the Indian e-commerce landscape, illustrating the power of technology and innovation in traditional business models. These success stories underscore the potential of technopreneurship to disrupt conventional business practices and drive economic growth. By embracing technology, these entrepreneurs have set new standards in their respective industries. Technopreneurship, the integration of technology into entrepreneurship, requires creativity, innovation, technical expertise, and a willingness to take risks. In an era where technology permeates every sector, technopreneurs are well-positioned to enhance productivity and improve business performance. By fostering a conducive environment for technopreneurship, governments, educational institutions, and industry stakeholders can ensure that technology-driven enterprises continue to contribute significantly to economic growth and societal advancement.

## 9.2 OBJECTIVES AND PURPOSE OF THE STUDY

The major objective of this research is to explore the realistic experience of the technopreneurs who have started their businesses in the city of Bangalore in the state of Karnataka. In this research four aspects are considered important which includes: competitive factors, technological and technical skills, innovation and creativity, and flexibility to adapt to new changes. Based on these factors the questionnaire has been designed (Rao et al. 2021). However, few exclusive questions have been framed to determine the various challenges and hurdles faced by these technopreneurs which includes financial issues, IPR issues, compliance issues and market competition. In addition, the secondary objective of the study is to explore the roadmap used by these technopreneurs to convert the challenges into opportunities and create business values (Verma et al. 2020). On the other hand, since most of the technopreneurs at some point of time would have applied for the loans and credits through the government schemes, this research also explores the applicability of PRISM, MUDRA and RKVY schemes introduced by the central government of India (Singh et al., 2020; Rao et al., 2020; Sharma et al. 2021).

## 9.3 REVIEW OF LITERATURE

In today's digital India, the contributions of the technopreneurs are enormous especially towards the growth and sustainability of the technical market. Not only the highly experienced technopreneurs but also young graduates have turned to be successful technopreneurs because of the vast experience and resources available to them at an easy reach. They have not just started business units which deal with technology related products and services but they have also provided some unique values through this delivery which has led to mass wealth creation for the nation (Kumar et al., 2022). Key players in the market play a crucial role in creating an impact on society and the potential for both wealth creation and societal benefit in the technology industry. Digital innovation significantly boosts GDP growth, business performance, and competitiveness, but entrepreneurs face challenges related to regulatory barriers, securing funding, and outdated technologies. Policymakers, entrepreneurs, and institutions need to collaborate to ensure that digital initiatives positively impact society on a larger scale and that the benefits are more evenly distributed. The relationship between digital innovation and entrepreneurship plays a crucial role in fostering business growth, increasing returns on investment, and promoting societal transformation (Kreiterling, 2023). The rise of technopreneurs has brought about significant technological advancements and innovations that have the potential to drive positive social impact. As technopreneurs, it is crucial to consider the ethical implications of their actions and innovations. Technology can help in economic growth and development' but at the same time it can bring huge disruptions to a nation as well. The major motto of any business should focus on societal contribution considering all the SDG's (N. Meade-Palmer, 2024). Hence, technopreneurs have to be change creators with positive thinking.

Technopreneurs face challenges such as access to operating capital, the absence of local markets, and concerns about risks in new businesses. Technopreneurs utilize high-tech methods to create new products and services, such as smartphones, social media, streaming videos, and cloud computing, showcasing their innovative capabilities. A strong desire for financial success drives entrepreneurs and focuses on the potential for success(Geek Tonight, 2023). Technopreneurship is an entrepreneurial activity in a technology-intensive environment, bringing significant economic benefits to society. The history of technopreneurship spans from the evolution of technology and its impact on entrepreneurship. Technopreneurs face challenges such as regulatory frameworks, cyber security issues, and skill gaps yet the main players in the market highlight the transformative potential of technopreneurship Atria University, 2020).

Technopreneurship combines technological innovation with entrepreneurial skills, focusing on collaboration among tech-savvy individuals to drive innovation and societal progress. Key sectors for technopreneurs include information technology, biotechnology, and high-tech products, intending to address contemporary challenges and foster economic growth. Technological entrepreneurship requires a blend of technical expertise and business awareness to introduce ground-breaking products and services, driving innovation in future industries (Sahadevan, 2024). The rise of side hustles has become a prominent trend in recent years, driven by individuals seeking additional income and pursuing their passions, further accelerated by the impact of the COVID-19 pandemic and economic uncertainties. Technological advancements, especially in cloud computing and AI, have played a crucial role in the growth and scaling of side hustles into full-fledged businesses, providing tools for small businesses to compete effectively and access global markets. A generational shift in work expectations, particularly among Millennial and Gen-Z, prioritizing flexibility, autonomy, and meaningful work, has led to entrepreneurship becoming an attractive option, supported by technology and the gig economy. The gig economy has experienced significant growth, offering opportunities for aspiring entrepreneurs to test business ideas, build a client base, and pursue their dreams, especially in the face of uncertainties reshaping the employment landscape. Side hustles, enabled by the gig economy and technological advancements, have become a valid form of employment, particularly preferred by younger individuals seeking financial stability, personal growth, and professional fulfilment, indicating a continued trend toward technopreneurship. Successful technopreneurs have founded tech companies that have made significant impacts in various industries, showcasing the importance of innovation and perseverance in technopreneurship. Start-ups play a significant role in the economy by driving innovation, creating new jobs, and disrupting markets. They transition into scaleups, which are more established companies with rapid growth and access to resources (Kollmann et al., 2022). The start-up in a country like India faced challenges a decade ago due to limited venture capital investment and underdeveloped start-up ecosystems. However, due to entrepreneurial motivation, the start-up ecosystem, strategic adaptability, and innovation capabilities are crucial factors in the success of start-ups transitioning to scaleups. Factors like financial barriers, non-financial resources, agility, and innovation play key roles in the growth and success of start-ups and scaleups. Hence, a strong and supportive startup ecosystem is being developed in the Indian market with the necessary legislative framework and a solid push from educational institutions to foster innovation and overcome growth-related challenges in start-ups (Crnogaj & Rus, 2023). Technopreneurs excel in innovation-disrupting norms, reshaping industries, and utilizing technology as a primary driver for their ventures. Technopreneurship significantly contributes to economic growth by creating new markets, generating job opportunities, stimulating demand, and benefiting the economy. Technopreneurs prioritize user-centric solutions, global connectivity, and the disruption of traditional industries, forcing them to adapt to remain competitive in the digital era. Individuals can leverage technopreneurship by embracing lifelong learning, identifying market gaps, fostering a growth mindset, collaborating and prioritizing user experience, and creating a social impact.

The COVID-19 pandemic accelerated the rise of technology-based start-ups, particularly in the realm of IoT and edge computing, leading to disruptive innovation and new business opportunities. The combination of IoT and edge computing has enabled the development of unified systems applicable across various sectors, from managing smart number plates to drone surveillance, signalling a transformative shift in entrepreneurship. Technopreneurs with foresight are leveraging edge computing and IoT to revolutionize industries, creating opportunities for new-age entrepreneurs to capitalize on the potential of these technologies. The youth in India are driving the establishment and maintenance of small businesses, leveraging technology as a key asset for entrepreneurial ventures. Implementing performance management systems can enhance employee performance and competitiveness in the industry. Small- and medium-scale enterprises (SMEs) need to address environmental uncertainty by effectively utilizing their workforce to mitigate risks and sustain their operations. Digital transformation is not just a trend, but a crucial opportunity for businesses to remain relevant and efficient in a rapidly changing tech landscape. Resistance to change, lack of clear vision, and skill gaps are key barriers that organizations face in their digital transformation journey. Successful digital transformation requires strategic planning, executive buy-in, and a commitment to learning and adapting to new technologies. Technopreneurship combines technological expertise with entrepreneurship skills to revolutionize traditional business practices and fulfill market demands. The surge in funding for tech-based start-ups, driven by the post-pandemic demand for technological solutions, presents a lucrative opportunity for aspiring technopreneurs. Successful technopreneurs have already proven the need and the opportunity to use technology to innovate and revolutionize existing industries, through these small-scale businesses also would scale up in terms of operations and strategies. Small businesses can foster innovation by adopting a beginner's mindset, challenging established norms, and exploring new possibilities without being constrained by conventional thinking. Intentional leadership and support for employees, customer-centric innovation, and routine integration of innovation into business activities are crucial for small businesses to drive growth and stay competitive. Embracing continuous learning, empowering employees to explore emerging technologies and industry trends can promote small businesses forward and open up new opportunities for innovation and success. The future of technopreneurship is vibrant, opportunistic, and driven by technological advancements through increased access to seed funding, easy global connectivity, strong ecosystems, and the importance of sustainability and social impact. Technopreneurs will play a pivotal role in shaping the future by solving complex problems and driving economic growth and development. However, there is a need for a structured framework and robust theory-building in the field of entrepreneurship, with a particular focus on technopreneurship. This would help in promoting the concept of technopreneurship from the root level of traditional business setup and also scale up the entire economy with technology integration.

In the ever-expanding field of academic study, a diverse range of perspectives come together to create a rich tapestry of insights into the complex world of Technopreneurship. At the heart of this narrative is the study by Maziriri and colleagues (2024), who delve into the critical role of attitude in shaping career trajectories within technology-driven professions. Their research unveils the intricate interplay between individual temperament, parental influence, and entrepreneurial drive, shedding light on the multifaceted factors influencing employment choices in the digital age (Maziriri et al., 2024). Meanwhile, Novani et al. (2023) set out to explore the dynamics of value co-creation among participants in the digital ecosystem. Utilizing a soft operational research lens, they meticulously examine stakeholder perceptions and interactions, providing practical insights to enhance cooperation and creativity within Indonesia's burgeoning ICT sector. In a related study, Kulkova et al. (2023) investigate the essence of technical entrepreneurship, emphasizing key characteristics such as innovation, proactivity, and risk-taking. Their research transcends traditional boundaries, offering profound insights into the transformative power of innovation within healthcare and beyond.

Muthusamy and Priyadharshini (2022) delve into the qualities distinguishing successful technopreneurs in Tiruchirappalli's software sector. Their findings highlight the value of applied creativity and problem-solving skills, painting a picture of perseverance and inventiveness in the face of challenges. Building on this foundation, Muthusamy (2020) defines technopreneurs as pioneers who leverage technology to spur corporate innovation and creativity, framing techno-entrepreneurship as a driver of economic growth and societal progress. Meanwhile, Bhat and Gulzar (2020) spotlight the obstacles faced by women entrepreneurs in the male-dominated technology industry. Their research underscores the necessity of addressing gender biases and creating supportive environments for women in technology. Gurana et al. (2022) offer a comprehensive overview of techno-entrepreneurship in India, emphasizing the contributions of leading technopreneurs to the national economy. Their research underscores the significance of innovation in addressing critical societal issues like unemployment, portraying techno-entrepreneurship as a vital driver of economic growth. There are a few factors leading to start-up success, focusing on education, experience, and resource availability. In addition, these factors provide crucial insights for aspiring technopreneurs, guiding them on the path to start up success.

Selvarani et al. (2022) characterize technopreneurs as risk-takers excelling in knowledge management and resource mobilization. It is very important to critically analyse the various perspectives on the evolving role of academia in the digital age, reimagining the ideal professor as a "technopreneurs" with a masculine tone. His research exposes crucial issues related to female representation and diversity within the technology sector. A little research work examines the disruptive impact of Taiwanese technopreneurs on China's industrial landscape, highlighting the interconnectedness of global technological networks . Some research works conducted in Kenya narrate the experience of the educated entrepreneurs, emphasizing the importance of knowledge sharing and collaboration in driving innovation and economic development. Throughout this narrative, Wibowo et al. (2023) explore the psychological factors influencing students' digital entrepreneurial aspirations, offering insights into the role of education and awareness in shaping the future generation of technopreneurs. Together, these diverse voices and perspectives weave a rich tapestry, illustrating the challenges and opportunities defining the realm of technology entrepreneurship. Each study adds a new dimension to the narrative, enhancing our collective understanding of this dynamic and evolving field.

## 9.4 PROBLEM STATEMENT

The reach of technopreneurs may sound very positive and vibrant when looking into the business success and the outcomes. Technopreneurs generate a huge income and earn name and fame in a short period of time. This is due to the creativity and uniqueness which they bring to the business. However, it is important to screen the various problems and real-time problems which they face in order to reach the peak position. Literature reviews convey a set of problems which technopreneurs have faced. However, these problems and challenges are not static; it keeps evolving from time to time in different scenarios. One of the biggest challenges is to be highly creative all the time and not only during the time of business initiation. In-addition, technopreneurs past life experiences, educational attainment, socio-economic status and other demographic factors also play an important role in their business success. Moreover, technopreneurs' role is not only to create wealth and maximize job opportunities for many potential employees but also to make huge contributions to economic growth in this "Digital Economy". Indian technopreneurs face a distinct set of challenges and opportunities that significantly impact their success in the fast-paced digital economy. One of the primary challenges is adapting to dynamic and uncertain markets, where new technologies are constantly emerging. This necessitates continuous innovation, requiring technopreneurs to stay highly knowledgeable and updated to effectively lead their teams and achieve desired outcomes. The need to remain current and sustainable in a knowledge-driven and innovative economy is a significant challenge (*McKinsey & Company, Simplilearn*). Moreover, financial constraints are a major hurdle, especially during the initial stages of setting up a business. Many technopreneurs lack substantial financial support early on, compelling them to work for other companies to gain knowledge and resources before fully venturing into their own businesses. This financial strain often leads to multiple failures before they can establish sustainable operations. Additionally, securing adequate funding is challenging due to the perceived high risks and regulatory complexities, which deters venture capital investment. Start-ups also face difficulties in acquiring and retaining skilled talent, exacerbated by an education system that does not always meet industry needs and intense competition from established firms. On the other hand, technopreneurs always have to be aware about the new creative ideas which are popping in the market creating an upset in the industry. Hence, one of the most crucial issues is to be updated with the intellectual property rights (IPR). These multifaceted issues collectively pose significant barriers to the success and sustainability of Indian technopreneurs.

## 9.5 RESEARCH METHODOLOGY

This research employs a qualitative approach where structured interviews are conducted with well-defined questions. This method allows for an in-depth exploration of individual experiences and provides detailed, context-rich insights into the specific problems technopreneurs encounter. Purposeful sampling was employed to select four technopreneurs from Bangalore. The selection of technopreneurs considered ten factors: diverse industry representation (e.g., Fintech, e-commerce, health tech) to capture a wide range of challenges, varying business sizes from early-stage start-ups to established companies, significant technological innovations, notable market impact, diverse funding sources including bootstrapped and venture capital-funded enterprises, substantial intellectual property portfolios, engagement with government programs, demonstrated growth trajectory and scalability, active contribution to the start-up ecosystem or entrepreneurial community, and receipt of industry recognition or awards for their contributions. A formal interview process was conducted using a set of 20 semi-structured questions designed to elicit detailed responses about the challenges faced by technopreneurs. This approach allows for flexibility in probing deeper into specific areas of interest based on the participants' responses. The interview guide was developed based on a comprehensive literature review and the study's objectives, covering themes such as financing challenges, intellectual property management, regulatory hurdles, market competition, and innovation strategies. Interviews were conducted face-to-face and using video conferencing to accommodate participants' availability. Each interview lasted approximately 60–90 minutes. Interviews were audio-recorded with participants' consent and later transcribed verbatim for analysis. Qualitative data from the interviews were systematically analysed using content coding techniques. A coding framework was developed based on the research objectives and key themes identified from the literature review. Initial codes were derived from the interview questions and refined during the analysis to capture emerging themes and patterns. Coding was conducted using a traditional method where an excel sheet has been initially used to feed all the important factors mentioned by the interviewee during the time of the interview. The following table (**Table 9.1**) shows the details of the technopreneurs who have been selected for the data collection. A purposeful and convenient sampling method has been employed to collect the data from them.

**TABLE 9.1**
**Showing the Technopreneurs Who Were Interviewed**

| Name of the Technopreneurs | Firms name |
|---|---|
| Mr. Bharath | Iotree minds LLP, Bangalore, Karnataka |
| Mr. Purav Joshy | KridaFan, Bangalore, Karnataka |
| Mr. Kishore B S | Fintrain Technologies Pvt. Ltd. (SMI – Stock Market Institute & Stockbyte App), Bangalore, Karnataka |
| Mr. Mohsin Hanif | Finclave, Bangalore, Karnataka |

### 9.5.1 Sampling Techniques

**Purposeful Sampling**: Find a selection of four technopreneurs based on predefined criteria including industry representation, business size, technological innovation, market impact, funding sources, intellectual property portfolios, government program engagement, growth trajectory, ecosystem contribution, and industry recognition.

### 9.5.2 Data Collection Procedures

**Development of Interview Guide**: Based on literature review and study objectives, a set of 20 semi-structured questions is created.

**Conducting Interviews**: Face-to-face and video conferencing interviews lasting 60–90 minutes, audio-recorded with participants' consent, and transcribed verbatim.

### 9.5.3 Data Analysis Process

**Content Coding Techniques**: Initial codes derived from interview questions, refined to capture emerging themes using a manual method.

**Thematic Analysis**: Identification and interpretation of significant patterns within the data, categorizing themes into areas such as financial challenges, intellectual property issues, regulatory compliance, and innovation.

### 9.5.4 Validation and Reliability Procedures

**Data Triangulation**: Ensuring validity by comparing data from different sources and perspectives.

**Member Checking**: Sharing preliminary findings with participants to verify accuracy and credibility. Once after the interview was over and after the transcription of the interview the output was shown to the technopreneurs to get their approval. This provides a pathway to attain the ethical process and procedures.

### 9.5.5 Reporting and Dissemination

**Compilation of Findings**: Organizing insights into a coherent narrative, emphasizing key success factors and challenges faced by the technopreneurs in the city of Bangalore. Every interview was carefully decoded to find the factors.

**Developing Recommendations**: Based on identified themes and patterns, practical recommendations for aspiring technopreneurs are formulated.

**Contribution to Academic Literature**: Findings are documented to enrich existing knowledge and provide a foundation for future research in Technopreneurship.

## 9.6 STUDY OUTCOMES

This study is to collect the realistic experiences of the technopreneurs who have faced many challenges in their technopreneurial journey. In addition, the study will help to know the various important key success factors which have inspired and motivated the technopreneurs to sustain and survive in the market. The driving force behind each and every technopreneur will add values to the future technopreneurs. Moreover, challenges faced by each and every technopreneur will differ based on the type of business, nature of market pressure, client expectations etc. Hence, deep introspection of the challenges will help to propose strategies for future entrepreneurs and technopreneurs to follow and benefit. Experience is always the best teacher. In this research interview is conducted to track few important factors such as: challenges, risk mitigation, regulatory framework, business motivation etc which adds more authenticity.

## 9.7 DISCUSSION

In these interviews with technopreneurs like Mr. Purav Joshy, Mr. Mohsin, Mr. Kishore, and Mr. Bharath, several common themes emerge. Successful technopreneurship emphasizes the critical importance of leveraging technology and continuous learning to navigate challenges such as regulatory compliance, financial hurdles, and market dynamics. They underscore the pivotal role of supportive networks, including family, friends, and mentors, in sustaining entrepreneurial ventures. Moreover, these innovators highlight the transformative impact of technological innovation on economic growth, illustrating how adaptive strategies and perseverance are essential in driving sustainable business success amidst evolving global markets. The discussions have been further explained below under various headings.

### 9.7.1 Common Success Factors

- **Technological Mastery and Innovation**: Across all interviews, a mastery of technology emerges as a critical success factor. Technopreneurs like Mr. Purav Joshy and Mr. Bharath emphasize the need to stay updated with technological trends to drive innovation and remain competitive. This includes adopting emerging technologies such as block chain (mentioned by Mr. Purav Joshy) and integrating new features into products (mentioned by Mr. Kishore).

- **Supportive Networks and Mentors**: Family support and mentorship play pivotal roles in entrepreneurial journeys. Mr. Purav Joshy highlights how friends and family provided crucial feedback and encouragement during the ideation phase. Similarly, Mr. Mohsin discusses the impact of learning from peers and mentor figures like Steve Jobs, indicating the importance of guidance and inspiration in overcoming challenges.

### 9.7.2 Challenges Faced

- **Regulatory and Financial Challenges**: Regulatory ambiguity, particularly in sectors like crypto and technology, poses significant challenges. Mr. Purav Joshy and Mr. Bharath mention navigating regulatory landscapes and seeking legal advice to ensure compliance. Financial hurdles, including initial funding rejections, were also prevalent, with strategies like diversifying funding sources (freelancing, corporate credits) proving crucial (as mentioned by Mr. Purav Joshy).
- **Educational Support and Skill Gaps**: The lack of formal entrepreneurial education in institutions was noted by Mr. Purav Joshy and Mr. Bharath. Despite supportive environments for technical skills (such as engineering labs), gaps in entrepreneurial training were evident, highlighting a need for curriculum adaptation to foster entrepreneurial mind-sets from an early stage.

### 9.7.3 Role of Government Initiatives

- **Limited Direct Impact**: While government initiatives like corporate credits (AWS, Google Cloud) provided tangible support (mentioned by Mr. Purav Joshy), technopreneurs like Mr. Mohsin and Mr. Bharath suggested that business growth often depends more on market traction and investor interest than direct government funding.

### 9.7.4 Impact on Economic Growth and Global Markets

- **Technological Advancements and Market Expansion**: The integration of technology not only enhances operational efficiency but also facilitates global market penetration (mentioned by Mr. Bharath). Mr. Kishore emphasizes how innovations in financial technology contribute to economic stability and global competitiveness, highlighting the transformative potential of technopreneurship in driving economic growth.

### 9.7.5 Continuous Learning and Adaptation

- **Lifelong Learning and Innovation**: Technopreneurship emphasizes the need for continuous learning to adapt to evolving market dynamics and technological advancements (as mentioned by Mr. Kishore and Mr. Bharath). This includes fostering a corporate culture that values innovation and employee development, ensuring sustained relevance and competitiveness in a rapidly changing landscape.

## 9.8 CONCLUSION

The interviews conducted with the technopreneurs in the city of Bangalore gave enormous amounts of information to trace the various problems they have faced in different stages of their technopreneurial journey. Technopreneurs survive in the most complex situation where every day is a challenge. The global technological upgrades are very haunting and have an impact even on the minute aspects of the technopreneurial business. The minds of the technopreneurs are always occupied with thoughts and ideas as their brain has to work continuously to deal with the new challenges in the national and international markets. At some point of time in their technopreneurial journey financial constraints reduce as they start generating much more than what is expected and super normal profits. However, dealing with new innovations, business synergies and to cope up with the new rules and regulations is very challenging. The Government of India should support the technopreneurs with more financial resources especially for the interested and capable aspirants. Existing successful technopreneurs should also provide training and development for the young aspiring entrepreneurs. This will lead to a win-win situation and benefit the society.

## REFERENCES

Atria University, (2024, September 22). Technopreneurship. Atria University. www.atriauniversity.edu.in/technopreneurship/

Bhat, I. S., & Gulzar, F. (2020). An insight into the factors affecting women technopreneurs. *International Journal of Knowledge Management and Practices*, *8*(1), 33–40.

Crnogaj, K., & Rus, M. (2023). From start to scale: Navigating innovation, entrepreneurial ecosystem, and strategic evolution. *Administrative Sciences*, *13*(254). https://doi.org/10.3390/admsci13120254

Department of Science and Technology. (2021). *PRISM (Promoting Innovations in Individuals, Start-ups and MSMEs)*. Government of India.

Geek Tonight. (2023). Technopreneurship. Retrieved from www.geektonight.com/technopreneurship.

Gurana, M., Kumar, D. S., & Choudhury, M. (2022). Empirical study on technopreneurship and technology-based companies in India. *International Journal of Innovative Science and Research Technology*, *7*(3), 1525–1538.

Kollmann, T., Kleine-Stegemann, L., de Cruppe, K., & Then-Bergh, C. (2022). Eras of digital entrepreneurship: Connecting the past, present, and future. *Business & Information Systems Engineering*, *64*(1), 15–31. https://doi.org/10.1007/s12599-021-00728-6

Kreiterling, C. Digital innovation and entrepreneurship: a review of challenges in competitive markets. *Journal of Innovation and Entrepreneurship*, *12*, 49 (2023). https://doi.org/10.1186/s13731-023-00320-0

Kulkova, I., Ivanova-Gongne, M., Bertello, A., Makkonen, H., Kulkova, J., Rohrbeck, R., & Ferraris, A. (2023). Technology entrepreneurship in healthcare: Challenges and opportunities for value creation. *Journal of Innovation & Knowledge*, *8*, 100365.

Kumar, A. (2022). Overcoming challenges in technopreneurship: Insights from Bangalore. *Business Today*. Retrieved from https://businesstoday.in

Maziriri, E. T., Nyagadza, B., & Chuchu, T. (2024). Innovation conviction, innovation mindset and innovation creed as precursors for the need for achievement and women's entrepreneurial success in South Africa: Entrepreneurial education as a moderator. *European Journal of Innovation Management*. https://doi.org/10.1108/EJIM-03-2022-0156

Meade Palmer, S. N. (2024). Technopreneurship and convergent thinking: Insights. Technopreneurship 360. https://technopreneurship360.com/2024/07/23/technopreneurship-and-convergent-thinking-insights-by-simon-n-meade-palmer/

Muthusamy, S. (2020). Evolution of technopreneurial entrepreneurs in the COVID-19 pandemic situation. *Research Explorer, 8,* 29–38.

Muthusamy, S., & Priyadharshini, J. (2022). Factors influencing the success of technopreneurship. *Analytical Commerce and Economics*, *4*, 31–45.

Novani, S., Cyntiawati, C., Kijima, K., Hasyimi, V., & Trianto, A. S. (2023). Empowering digital creative ecosystem using problem structuring methods and a service science perspective: A case study in Cimahi and Bandung, Indonesia. *Asia Pacific Management Review*, *28*, 215–228.

PMMY. (2021). *Pradhan Mantri Mudra Yojana*. Government of India. Retrieved from https://india.gov.in

Rao, P. (2020). Continuous professional development for technopreneurs: Strategies for success. *Entrepreneur India*. Retrieved from https://entrepreneur.com

Rao, S. (2021). Integrating technology with entrepreneurship: A pathway to economic growth. *YourStory*. Retrieved from https://yourstory.com

Sahadevan, S. (2024). Exploring the challenges and business opportunities in technopreneurship. *Journal of Emerging Technologies and Innovative Research (JETIR)*, *11*(4), b669–b677. www.jetir.org/

Selvarani, A., Venusamy, K., & Devrath, R. C. (2022). Entrepreneurship and technical approach: Success stories of technopreneurs. *International Journal of Advances in Engineering and Management (IJAEM)*, *4*(8), 76–81.

Sharma, N. (2021). Government initiatives fostering technopreneurship: An overview. *Indian Express*. Retrieved from https://indianexpress.com

Singh, R. (2020). The importance of incubation workshops and technical sessions for technopreneurs. *Inc42*. Retrieved from https://inc42.com

Verma, T. (2020). Navigating the challenges of technopreneurship in the Indian market. *Times of India*. Retrieved from https://timesofindia.indiatimes.com/blogs/voices/navigating-2023-what-entrepreneurs-and-start-ups-should-be-mindful-of-in-the-new-year/.

Wibowo, A., Narmaditya, B. S., Saptono, A., Effendi, M. S., Mukhtar, S., & Shafiai, M. H. M. (2023). Does digital entrepreneurship education matter for students' digital entrepreneurial intentions? The mediating role of entrepreneurial alertness. *Cogent Education*, 10(1), 2221164. https://doi.org/10.1080/2331186X.2023.2221164

WIPO (2020). World Intellectual Property Indicators 2020. Geneva: World Intellectual Property Organization. ISBN: 978-92-805-3201-2. Available at: www.wipo.int/ipstats.

## Annexure

### Interview Schedule

Critical Success Factors:

- What do you believe are the critical success factors for technopreneurs in India?

Challenges Faced:

- What are the primary challenges you face as a technopreneurs?

Role of Government Initiatives:

- How have government initiatives such as PRISM and MUDRA loans supported your entrepreneurial journey?

Market Opportunities:

- How do you identify and exploit market opportunities in the digital age?

Intellectual Property Rights (IPR):

- How do you manage your intellectual property rights, and what challenges have you encountered in this process?

Integration of Technology and Business:

- How do you leverage technological advancements to create business opportunities?

Role of Higher Educational Institutions:

- How have universities and higher educational institutions supported your journey as a technopreneurs?

Innovation and Creativity:

- How do you foster innovation and creativity within your organization?

Financial Challenges:

- How have you secured financing for your startup, and what challenges have you faced in this regard?

Regulatory Compliance:

- What regulatory challenges have you encountered, and how have you addressed them?

Impact of Technopreneurship:

- How do you believe technopreneurship contributes to economic growth and societal advancement?

Case Studies and Experiences:

- Can you describe a significant success story from your journey as a technopreneurs?

Support Programs and Training:

- How have professional development programs and training workshops influenced your success?

Corporate Culture and Innovation:

- How do you cultivate a corporate culture that rewards innovation?

Global Market Opportunities:

- How do you approach expanding your business into global markets?

Role Models and Inspiration:

- Who are some technopreneurs that have inspired you, and why?

Economic and Societal Impact:

- In what ways has your venture addressed specific societal challenges such as healthcare or education?

Continuous Learning and Adaptation:

- How do you ensure continuous learning and adaptation in your business practices?

Balancing Technology and Tradition:

- How do you balance technological innovation with traditional business practices?

Future of Technopreneurship:

- What do you envision for the future of technopreneurship in India?

## Interview transcriptions for reference

### Interview 1 with Mr. Purav Joshy

#### Success Factors

Meeting good friends, awareness about Shark Tank, ideation, sharing the idea with friends, and getting their feedback. He said family support is equally important, without which, it is difficult to move forward. He has a good co-founder with 7–8 years of experience.

### Education and Business Success

Technology is the buzzword. He did his engineering in 2012 from Pune. Initially, he worked at many small start-ups and L&T, then came to Bangalore but always had dreams that he would have a start-up of his own. It was an accidental entrepreneurship. He was clear from his college days that he would be starting an outsourcing company. He was introduced to block chain technology in 2017. He used to read a lot of vlogs, newspapers, and other online materials rather than physical books. Life is a cycle with ups and downs. Then, he found a group of friends who were also active. Collectively, they started exploring this industry. He was influenced by his friends. Initially trying to understand his interest and inclination, COVID-19 IPL drew his attention. In 2021, he started with ideation and shared his ideas with his friends because was fantasy football with a niche space.

### Challenges Faced

He did not have any guidance. Mentors or mentorship makes a lot of difference. They did not know the positioning or how to identify the target users. They got to know maybe after 6 months. It was like they were getting the parts while flying and assembling them while flying. So, when he asked another founder, they said it is the same across start-ups. There is a difference between crypto start-ups and non-crypto start-ups. He said he did not get proper guidance. They had to "recreate the wheel".

### Government Schemes

They are not a pure web-based tool, so they could not avail any benefits from government funding. However, he got benefits from corporate credits. Various corporate credits like AWS credit, Google Cloud credit, and MongoDB Credits for start-ups. It really helped at a stage when they were trying to get started, and it elevated their efforts.

### Exploring Market Opportunities

Marketing opportunities were explored through need positioning and target positioning. Positioning geographically: Needing to find where there were potential football fans. They explored the market outside India. Football is popular in Goa, West Bengal, and Bangalore, which is also a hub for football and crypto. If there was any conference anywhere with target users, they would attend to identify the pulse of the market.

### Educational Institutions' Support to Entrepreneurship

There was not much support from educational institutions because mobile phone use was in its infancy around 2012–2014 when the boom happened. He did not have a start-up culture or platform in his university. He learned a lot from newspapers, particularly Business Standards. Reading from the front page to the last helped him understand the terms. Every Monday, they have four pages on start-ups. Institutional support in the form of grooming and teaching soft skills like communication and psychology should be taught at higher school levels. He referred to the example of Mr. Warren Buffet, who started a business at the age of 12, and how he learned.

### Innovations

Technology is something that keeps every entrepreneur on their toes. Therefore, to survive, they have to be innovative and adapt to technology. They provided a support system for employees to learn what they would want to learn.

### Support Programs and Training

When they employ a person, they ensure that the person has the ability to learn. Both co-founders have rich experience. They ask them why they want to learn, push them to explore, and guide them with different opportunities to learn.

### Regulatory Challenges

Formation of the company, because it involves fantasy football with crypto, was a challenge. Regulation in India is not clear for crypto start-ups. They are not a pure web tool, and it is not under the purview of the Indian government. Regarding legal aspects, they have their Chartered Accountants who will take care.

### Financial Challenges

They faced a lot of challenges in raising funds. They almost faced 30–35 straightforward rejections. At one point, they even thought of giving up. But then, they strategized so that one person focused on raising funds while the other co-founder focused on building and crafting the product. In fact, none of the VCs from India helped; large VCs like Polygon Venture Fund, Shima Capital, and others.. They took the rejections as opportunities to answer better next time.

### Impact of Technopreneurship: Contribution to Economic Growth

Technology is the buzzword. All entrepreneurs will become technopreneurship one day. That is the pace at which technology is moving. Therefore, technopreneurship' contribution to economic growth is huge because they are going to give exponential growth in productivity.

### Global Market Opportunities

It is fantasy football with crypto space, so the opportunities across the globe are huge. There are many football fans throughout the world. Hence, the global market is wide. The only thing is to identify and capture the market.

### Experiences

Conduct a user interview with your product, let it be alpha, let it be beta. So, when we started, we had our own hypotheses. We used to put X on the block chain and Y on the block chain, and the user would cross-verify, and then the user would think it cannot be modified. We have revised our hypotheses based on the user because they are ultimately going to use the interface. First, when we started, we had a 3D ground, and when you pick a player, it was like animation. then they used to say they liked 2D better because they have been using 2D for a long time, maybe for 15–20 years, and I am asking them to change their habits.

**Role Models and Inspiration**

He learned a lot from his colleague at L&T. He got inspired by APJ Abdul Kalam, and during 2013 and 2014, Elon Musk was in full swing. Their ideas and the way they give speeches. Then Steve Bob.

## Interview 2: Mr. Moshin

**Critical Success Factors:** The one critical factor is that you should learn technology first. Otherwise, there are cases where you can get duped when you are dealing with other tech people with the know-how in the market. Because these days, you often find people working on multiple tasks or multiple companies. So, it's very difficult to judge how much time a project can get completed or a task can get completed.

**Challenges Faced:** Development of tech is the major challenge that we have faced. Ongoing changes in technology is something that poses a huge risk as well. Basically, development of technology, ongoing changes, costs of servers and other things. My belief is that intellectual properties are there. We are trying to get it copyrighted in many ways. But unless and until whatever you develop, unless and until what I personally believe is, It's easy for a person to copy your intellectual property, but he will not be able to implement it in the way that you are implementing it. So, copying intellectual property is one thing, but implementing it is another thing. Because it's your intellectual property, you know how to do it better.

**Role of Government Initiatives:** Basically, the thing is I had tried for various funding from the government, but I could not secure it. Maybe because I could not concentrate on the steps that had to be followed at that time. And because of multiple things in hand, I have never been in a situation where I have approached anything related to Prism or Mudra loans. Yes, government support has to be there. But ultimately, what I have realized is from a change of what I was before 10 years and what I am today, it is all about how you are able to grow your business. If you are able to grow your business in the right way, if not the government, there are a lot of investors sitting out there in the market who will certainly support you. So, it is not about the support that the government could give. Obviously, the government is giving a 20 lakh grant and 80 lakh debt funding at a rate of 4% or 6% in the market currently. But at the same time, I feel that is free money. Again, there are a lot of things there in the market wherein a lot of bribes and everything is going on. But yes, that is free money. If someone is able to get it, you get it. But for me, what I personally feel is, be it government support or whatever it is, if you are able to do your product in the right way, you will have 100 investors lined up who would want to invest in your product. If your product or if your service or whatever it is, is not good, you will not get investors. Additional support in terms of, as I said, yes, the government has been doing a lot of good things out there in the market.

**Market Opportunities:** Basically, I am into financial markets. One way which I can tell is one way how I've had the society is that if you are an investor or trader in the market, there is always greed, there is always fear out there when you are doing your first investments or trading or investing in the market. I have just tried to remove greed and fear from the markets for my clients. That's what I've given to society. And also, people who are not able to adapt to me, we do not work with them.

**Intellectual Property Rights (IPR):** Currently, we are in a company called Finclay. So yes, this is a word, this is a term that has been used by a couple of colleges for their Fests/events. And recently in the year 2021, a bigger event for influencers has happened. And what we best can do is we can send them notices that, okay, you're not allowed to use this name because I own the copyright. I own the trademark of that particular name. And then maybe they come to us with some barter, exchange or they come to us with some requests… .you have to allow us to do this because we have already done a lot. So, X, Y, Z things happen. And then at times you allow them to use the name. At times we do not. We have had around 17 cases till date.

Again, securing our name, securing our brand, securing a couple of terms that we use in terms of marketing. So that has helped us. If someone is using it, we can always send them a notice. And again, in terms of patents, we are just trying to get a few patents. So, we do not have any experience in terms of patents.

**Integration of Technology and Business:** Again, I would say the same thing that in our field, technological advancement is equivalent to nothing but speed as well as cost of saving data. These are two things that are always a hurdle. So, we keep upgrading our technology as and when as and when required. And whenever new technology comes, if we feel that's right, if we feel that's cost effective, we upgrade it. And that's how new business opportunities are created. Say, for example, Algo trading in India was not so famous. It started capturing fame in the year 2020 during the lockdown days. And at that time, we did not have good technology to do Algo trading. Currently, we have better technologies in terms of Algo trading, in terms of speed, in terms of servers. So as and when we feel that, OK, there's a business opportunity and we can always do better in terms of using some technology, we always do it.

**Role of Higher Educational Institutions:** Yeah, I would always believe in two things. If at all a person who is graduating out of college wants as a target of just doing the job for his own life, he should at least be taught a little bit of finances so that when he goes out, he knows how to manage his finances. Additionally, I feel if at all a person who just wants to go out of college and do a job, he would obviously require the use of technology in his daily life. So, he should know a little bit of technology. So, hand in hand, I would say that finance and technology should be included in all courses so that a person graduating out of college knows finance and technology so that he is successful in doing what he wants to do out there in the market.

**Innovation and Creativity:** Innovation comes in with experiences of your own self with the business, with experiences of you and your client or you and your customer. A customer is facing X, Y, Z issues. And you think that, OK, if we are going to implement the issue in this way, in this innovative way, the customer will not face such an issue, and the customer is going to be happy. That's how we keep innovating and innovating in our organization. As per the needs and requirements. One thing we always have to take care of is we are a semi registered entity. So, we always have to take care of compliance. So, whatever we do, we need to ensure that our compliances are in place. And we also adapt to the changing technologies. So, on one side, it is compliance and on the other side, technology. I think together we need to constantly upgrade and update ourselves so that we are within the compliance aspect.

**Financial Challenges:** In the initial start, I was actually looking out for finances, and I have approached a lot of people wherein I could not secure any kind of finances. And later I realized that, OK, I am not doing something right. That's what is something that's stopping me from getting finances in the market. And I'm not able to convince investors because I am doing something wrong or I'm not doing something that's going to be successful in the future. That's why investors are rejecting me. So I sat across and understood what could be done so that my business runs, runs better. That's when the investors are going to invest. So when I sat across and when I started implementing my business in a better way, started improving on the technology, started improving on my sales, marketing, getting in artificial intelligence everywhere, getting in machine learning everywhere, into our trading, into our investments, our business automatically started growing. And currently we are a team of around 90 plus people and we are not in need of any funding assets. Maybe in case we advance in the future, in case we grow in the future, maybe we will be requiring funding. But currently what we do is we always keep around 30 percent as reserves so that it can be used for any further project that we are doing.

**Regulatory Compliance:** We face a lot of compliance aspects, starting from at times even compliance from the investors, which we have faced around three complaints from the investors through the SEBI route. There are a lot of wrong things that are happening in the market in terms of compliance wherein we can get stuck. So, we are ensuring that all the agreements, all the disclosures, all the steps that a client has to do are all very clear. To stay informed we go into the extent of reading daily emails that SEBI sends. Every day there are 10 cases that SEBI gives 10 to 15 cases that SEBI gives, or SEBI gives its decision on. Say, for example, someone has made a mistake out there in the market or someone is not compliant with XYZ rules and regulations. So SEBI gives out a case. Sometimes it's a hefty fine. Sometimes it bars people from attending the people from attending the markets. So, we sit and read all those cases. We spend at least 40 minutes every day reading all these cases as to what's happening there in the market. And how can we save ourselves from these things that are affecting others? So that's how we keep ourselves updated on any upcoming regulations. We keep reading them and we think we keep implementing technology there as to how we can be extra compliant instead of being compliant in our business. That's what keeps us going.

**Impact of Technopreneurship:** Technology is the next big thing and technology is everything. If you have to ask me in terms of our financial markets, the current generation or the future generation will not have so much, will not be able to manage so much stress which the older generation used to face in the markets. So, in my field, unless and until there's no technology that's implemented, it's very difficult to go forward. Considering this is a field where there is always stress involved, where there's fear involved, where there's greed involved, which can make a person unsuccessful.

**Case Studies and Experiences:** So, as I started by this by saying the fact that if you do not know technology, you cannot be successful. And if you cannot be successful in a business related to technology, the same thing goes here. When I started my journey into doing something related to technology, stock market, I was never successful because I almost got duped by around 4–5 developers or development companies out there in the market who promised me to develop certain things and give me. But ultimately, I realized in the year 2020 that unless and until I do not learn, I will not be able to delegate tasks to someone. So, I started learning. I started doing courses. And that's when I could hire people and that's when I could ask those people to develop what was required for me. I understand that the other way is to hire a tech director or whatever it is or a tech founder. But I believe that I believe that unless and until I do not know, the other person can always dupe me. So, I learned it and that's how we are going forward.

**Support Programs and Training:** Frankly speaking, I've attended many. It has benefited in terms of what problems others have faced and how they have been able to succeed out of their problems. That's what has helped me more than attending the workshop to do this, do that. What I personally feel is that I as a person do not learn from workshops, but me as a person learns more, by understanding others' cases and what others are doing and how others came out of issues rather than me learning from a trainer. I understand learning from a trainer is very important, but that has never helped me. Case studies inside a workshop have helped me more.

If you have to ask me, it's about how people have to be taught about how to handle business better, how to organize yourself better, how to do time management better.

**Corporate Culture and Innovation:** In terms of corporate culture, we have a good culture. Everything is about appreciation. So, if you do not appreciate the other person, the other person is not going to feel good in the organization and then the organization will not grow. Everything is about appreciation and support and less politics. So, these are the three things that we follow: appreciation and support and less politics.

**Global Market Opportunities:** I do not see any challenges entering the international markets unless and until the challenges would be how do you how do you enter there? How do you set up your business there? Setting up a business is always going to be a challenge in the international markets. Setting up a team in the country that you're targeting. But that's what I would see as a challenge. Apart from that, because you have XYZ processes that are set in India, I think these processes are going to be obviously beneficial there. So, I do not think anything new is needed. Okay.

**Role Models and Inspiration:** I do not have anything as specific as a role model at all. I am a person who would just listen to everybody. And I would learn from everybody what others do and how I can implement that better. So that's what I do. I do not follow anybody specifically. So that's what I do. Factors: No specific role model, I learn from everybody and see how I can implement that better (Continuous learning from surroundings).

**Economic and Societal Impact:** Nothing in terms of education. We don't have any measure for societal impact. We have taken out greed and fear from the markets for our clients.

**Continuous Learning and Adaptation:** Customers help me understand my limitations and I can work on them. Factors: Learning from people, surroundings, others' mistakes.

**Balancing Technology and Tradition:** It is all about combining both. Traditional business practices are automated. Not faced cultural or traditional challenges. Small things happen here and there, and we ignore them. Factors: traditional practices are automated. Ignoring small cultural and traditional challenges

**Future of Technopreneurship:** It is all about technology. The future is going to be digital. We have to be digital; we have to get ready for it. To innovate, we look around, what competitors are doing, understand what the problems are faced by the clients and our own problems in business and how to overcome those Factors: Digital future, Preparation for the same. Understanding competitors, clients' and our own problems and to overcome that we innovate

## Interview 3 with Mr. Kishore

### Impact of Technopreneurship

It certainly contributes to economic growth. Technology is inevitable. Our venture has offered educational programs, after assessments are certified. One specific example is only when they have come back and told us that we have realized the impact we have created. For example, IDLC(Bangladesh), after approaching many and not being satisfied, came to us and we catered to their requirements by customizing as per their requirements. They sent their ppl to get trained with us. There has been direct and indirect impact.

### Case Studies and Experiences

There have been many significant success stories, for example one student wanted to enrol only with us, after seeing that his friend had bought a house after doing a course with us. Also about a family, where the father passed away and the family was asking for money from friends and relatives and his son when he did a course with us went back and checked the details and realised that there was an insurance policy which they were not aware of and did not know how to go about it. After learning through the course, my son has understood and set things right and they are doing well now. Many times, when others copy us etc., we feel bad, but one thing that we maintain is doing things right and believe in ourselves, work hard etc.

### Support Programs and Training

I keep attending sessions and workshops other than finance/stock market related as I will get perspectives from different domains as to what it is that they get or don't get in finance. Apart from that every day, six of us, no matter where we are in the city, we reach office for lunch and discuss and deliberate about various topics finance and non-finance and learn from others mistakes and learn from success stories. Additional support in terms of funding, understanding new things that we are trying to pitch in, what we are offering etc. will be beneficial.

**Corporate Culture and Innovation**

We keep killing our existing products and keep bringing out new things with more added values. Constantly innovating and offering what is needed is something we keep doing. We deliberate and constantly take feedback and inputs from others as to in what way we are doing good. What if it was there was feedback, it would have been good? What are the customer's needs and how can we customize as per their needs.

**Global Market Opportunities**

Compliances could be a challenge. However, in India we have more compliances compared to foreign markets. We have all kinds of features in our products. As a technopreneur we are updated about the latest laws and regulations in the country.

**Role Models and Inspiration**

Steve Jobs and other big names are an inspiration. However, we constantly learn from people around us. We take their inputs and feedback. We constantly listen to others about what they have to say about our products. Also, we six of us, our everyday discussions over lunch, keeps us always charged (food for body and mind together).

**Economic and Societal Impact**

We obviously look deep on strategies and work on enabling and empowering women in finance and financially educating them to make their financial decisions, be part of financial decision making in the family. We definitely feel this initiative made a difference to women and empowered them. Through our business we have extended many opportunities for women employees.

**Continuous Learning and Adaptation**

Staying updated, not just about our own field but about everything around, discussions everyday over lunch and constant innovation. But we also need to know what is happening in the global context. If we are not aware about what is happening in the technological world our business will suffer.

**Balancing Technology and Tradition**

In the initial years we have faced the challenges in terms of people not believing in what we have to offer, not investing, thinking that we may misuse the data etc, but as we have evolved, these challenges have greatly eased out.

**Future of Technopreneurship**

Technopreneurship will grow as technology is inevitable. We wish they get enough support and trust when they speak about what they have to offer. We kill our products and keep on innovating. We take feedback, inputs from others ongoingly and that's how we are able to incorporate numerous features in our products. Our competitors have benchmarked us.

## Interview 4 with Mr. Bharath

### Education and Business Success

Technology can enhance business revenue, leveraging skills from a hardware and IoT background to expand from domestic to global markets. Implementation of technology is guided by past experience in overcoming hurdles and providing solutions.

### Government Schemes

Initial funding obstacles were overcome by starting services and freelancing to fund business dreams. This preceded registering the business and setting up a proper business structure, which later enabled leveraging government policies.

### Exploring Market Opportunities

Business ventures aimed at scaling revenues for expansions. Understanding client expectations, reducing operational costs, and enhancing efficiency are crucial, akin to Toyota's focus on internal system functionality and operational correctness for revenue enhancement.

### Intellectual Property Rights (IPR)

IPR is a significant issue in India, especially for technology-based entrepreneurs. Research-based products require quick conversion from products to revenue, intensifying competition around IPR.

### Annexure-2: Educational Institutions Support to Entrepreneurship

Despite educational constraints in curriculum structure, support from engineering faculty provided essential lab facilities for experiments, though the curriculum itself did not adequately support entrepreneurial journeys.

- **Innovations:** We train 25 to 30 employees constantly. So, in this regard we as trainers to impart the best technical knowledge have to be equipped with advanced global skills. We ensure to keep ourselves updated. This also helps us to be more creative, innovative and provides a motivation to look ahead.
- **Service-Based Companies and Client Expectations:** Client firms' expectations are huge. Each and every client firm sets a different benchmark. Hence, as technopreneurs we ensure that we deliver the best values to all our clients and other associated parties.
- **Regulatory Challenges:** Regulatory changes are quite often in the technopreneurial landscape. Hence, we ensure that we are updated and also know what the latest updates in policies.
- **Technopreneurship Adding Value to Business Arena:** The first business value which we create is to provide products and services with added values as per the needs of our clients. Secondly, we strategize and come out with many new strategies which not only helps our business but also impacts the associated parties. Hence, the entire business ecosystem is empowered.

- **Support to Employees – Training:** We have a set of dedicated senior managers who take up the keen responsibility to train the juniors. Five days' work and Saturdays and Sundays are provided for the wellbeing of the employees. Because employees are our priority, and they are the most valued assets.
- **Corporate Culture and Employees:** In the technopreneurial landscape being very competitive is the most important aspect. In addition, we have a strong learning culture where each and every employee actively learns and benefits.
- **Technology and Global Operations:** First and foremost, the technological changes affect the global operations. Apart from these financial aspects, inflation, new policies and changing client needs and customer needs are the major factors affecting the global operations.
- **Role Model and Inspiration:** The TATA which has spread across the globe is the role model for business and for me. The foundations are so strong that the brand has always grown and established itself. Moreover, it is known for technological advancements.
- **Social Issues:** Technology plays a major role not only in the business world but in the life of everyone. Hence, social issues can also be solved using technology. Business should focus on how technology can bring a revolution in a positive manner in the country helping the life of many.
- **Lifelong Learning:** Technopreneurs have to be involved in constant learning and development. The world's technology develops every day and hence to cope-up with this it is important to be updated.
- **Future Prospects of Technopreneurship:** More than just entrepreneurs, technopreneurs have to be more creative and more sound in technology. Because the technical industry / sector is changing rapidly. There are good opportunities in technopreneurship.
- **How to Plan to Be Innovative and Creative:** Maintain a clear vision and ensure to provide the solution as per the needs and requirements. Ensuring the value addition in every stage and active involvement to adapt to change is very important.

# Index

## T



## V



## W